Careers with a Science Degree

Revised by
Eileen De'Ath and Tessa Doe

based on the previous edition
by Mary Munro

Student Helpbook Series

11506

Careers with a Science Degree - 2nd edition September 1999.
Produced in association with UCAS.
This edition has been revised by Eileen De'Ath and Tessa Doe
based on the first edition written by Mary Munro.

Published by Lifetime Careers Publishing, 7 Ascot Court,
White Horse Business Park, Trowbridge BA14 0XA

© Lifetime Careers Wiltshire Ltd, 1999

ISBN 1 873408 93 5

No part of this publication may be copied or reproduced, stored in a retrieval system or transmitted in any form or by any means electronic or mechanical or by photocopying or recording without prior permission of the publishers.

Printed and bound by Arrowhead Books, Reading

Cover design by Jane Norman
Text design by Murray Marshall
Cover photograph used with the kind permission of the University of Bath.

Contents

	Acknowledgements	*5*
1	*Why science?*	*7*
2	*The image of science and scientists*	*19*
3	*Women and men in science*	*37*
4	*Science courses in higher education*	*55*
5	*Choosing and working with your science degree*	*83*
6	*Where will your science degree lead?*	*97*
7	*Postgraduate study - what are the options?*	*113*
8	*International opportunities*	*131*
9	*Your future as a science graduate*	*143*
10	*Where do you go from here?*	*153*
	Book list	*163*
	Glossary of science courses	*167*

Acknowledgements

Thanks are due to the following for assistance in revising and updating the information for this second edition of *Careers with a Science Degree:*

Tony McQuaid of the Qualifications and Curriculum Authority
Andy Davidson of the Institute of Employment Studies
Tracy Brennan of the Association of Graduate Recruiters
Richard Wiltshire of Lifetime Careers Wiltshire Ltd
Dr Jan Peters of the Promoting SET for Women Unit
Paul Raymond - Editor of 'What do Graduates Do?'
Lynne Puntis and pupils of Aloeric School, Melksham
Sam Kearsley of UCAS
Caroline McGrath of the Association for Science Education
The Royal Society
and
Andrew Smith of Smith Owen Associates for providing the glossary.

Thanks also to Rob Brown, Paul Compton, Jane Davies, Rebecca Dumbrell, Jessica Harris, Gary Hepherd, Celia Keen, Mike Partridge, Paul Pilkington and Sue Wood for providing us with their career profiles, and to Jacki Ciereszko for her DTP skills.

Please note that we have tried to incorporate relevant information where appropriate about the forthcoming changes to post-16 qualifications in England and Wales and in Scotland. How these changes actually work in practice is yet to become clear; please seek more detailed up-to-date information.

Chapter 1: Why science?

'School science provides pupils with a knowledge of the natural world, the skills of investigation and experimentation and an appreciation of the importance of science to individuals and society. Science also develops such personal skills as curiosity, motivation, teamwork and the ability to communicate. These are widely recognised as particularly important skills and values, both in science and as part of a broader education. They help prepare pupils for further study and for a broad range of careers, as well as providing a basis for informed citizenship.'

Statement from the Royal Society, 1998

A valuable tool

Scientific understanding is of great value - to individuals and to society as a whole - in many areas of modern life. Whether our interest lies in the development of groundbreaking scientific ideas and new products, concern for our impact on the environment, crucial issues of health and safety and world health problems, or a natural curiosity about what things are made of and how they work, science is important to us all.

Taking a science degree does not mean that you are choosing to spend the rest of your life working in a laboratory. Many science graduates want to be lab-based scientists, but there are lots of others who use their science degree as a stepping stone to other careers. Science graduates can go into most things that other graduates do, and many more. They can consider job areas where a science degree is required, as well as careers where any degree subject gains entry.

Nowadays we have to be prepared to change direction several times in our working lives. A science degree gives you the most options to choose from. Many of the career and postgraduate study opportunities are open to holders of Higher National Diplomas as well as to graduates.

What this book is about

This book will help you to connect science subjects to the wide range of job choices open to science graduates. It encourages you to reflect upon some of the stereotypical images of science and scientists in Britain, to look beyond the images of boffins in white coats with limited career choices to the real challenges and opportunities that science offers. It takes a look at science globally, and helps you to understand the fast-changing world of graduate employment.

The subject choices you make at school have an immediate effect on the range of jobs that are open to you. If dropped after GCSE, science subjects are difficult to take up again at a later stage. Too often, when people come to apply for higher education they find their choices are limited by decisions they made when they were much younger. That is why this book is important to you now.

Keep your options open

- Find out about the wide range of careers open to science graduates.
- Avoid shutting off any career routes which you may later want to reconsider.
- Try to allow enough flexibility to enable you to develop new interests and skills.

Science is different

Why does this book focus on the sciences rather than all the other subjects people study at school and college? Why is science different?

- Science is about concepts and ideas. Your knowledge and understanding of science builds up gradually. Understanding new concepts often means building on other concepts you have already learned and understood. This is why you cannot go on to science A level, Advanced GNVQ or BTEC National courses without studying sciences at GCSE. In Scotland you would not take a science for Highers or Advanced Highers unless you had already done well in the subject.
- Scientific concepts are much easier to learn when you are younger. This is particularly true in mathematics: the essential tool of all science. But it also applies to physics, chemistry and biology.

Science is for the young

As you get older, it is much easier to move from jobs that require a high level of scientific knowledge into those that require other knowledge, skills and experience, than it is to move in the opposite direction.

Many people with science degrees find that they become gradually less concerned with new scientific concepts as their careers progress. Many go on to take further qualifications in other things, such as financial or human resource management, marketing and sales or technical writing. Of course there are those in research and academic jobs who will continue to work at the frontiers of science throughout their working lives, but even they probably do their most creative work when they are younger.

So, if you are going to study science either for a career or as a basis on which to build other skills, now is the time to do it! Take some time to think about it seriously.

Don't miss the boat!

Think about a future in science now. It may be your last chance to do so. Once you have stopped studying science it is very difficult to go back to it. If you don't take science at A level or equivalent it is harder to get into science in higher education. And if you don't do a degree or HND course in science you are very unlikely to be able to follow a career in science or take a science-related postgraduate course. This is quite different from other areas like business or accountancy, where many people with degrees in other subjects (including the sciences) go on to successful careers.

The decision to drop science is very hard to reverse later on. It's a sort of intellectual one-way escalator!

Looking beyond Frankenstein

Yet another reason why science is different is because the subject at present carries many negative images and stereotypes, particularly in Britain. A stereotype is an image that is fixed and is not changed by discussion or persuasion. You may find a grain of truth in some stereotypes, but it is usually buried under layers of ignorance and prejudice.

People have preconceived ideas of science being difficult, boring, inhuman, laboratory-based and - despite decades of equal opportunities - a career for men. The image that lingers of Doctor Frankenstein as an isolated (male) manufacturer of monsters is very powerful. The debate about genetically modified foods, cloning and 'scientists playing God' has often evoked this stereotype.

Dr Frankenstein is a fictional character created in 1818 by Mary Shelley. He is an idealistic young scientist who finds the secret of giving life to matter and creates a living being from parts of dead bodies. The monstrous creature is so feared by those who meet him that he becomes embittered and cruel and turns against his creator.

Mary Shelley's story has shaped one of the images of scientists in our society. It is extraordinary, in spite of our increased knowledge and awareness, that this mythical man has remained such a powerful influence for so long.

There is also a very close association in many people's minds between science the subject and science the career. This is not so in other subjects. Those choosing to study history are seldom influenced by their image of a historian! Few students of French imagine themselves as French professors or interpreters. People usually choose arts and humanities subjects because they are interested in them or because they are good at them, not necessarily because they lead to a particular career. Why should science be any different?

We should think of science as an important part of our general education, as it is considered in many other countries. And if you wish to use your scientific knowledge and skills in your career, there are many other areas besides scientific research where you can do so.

Chapter 2 of this book is about the image of science and scientists, where these perceptions come from and how they affect subject choice.

Is science difficult?

When people miss out on a few classes for some reason, or just come up against something they find difficult to understand, they often become discouraged and lose interest. This is particularly true of science subjects.

You may find the concepts and ideas in one science subject harder than in another. Everyone is different in the way they approach subjects. There are people who find maths and physics so easy that they can't see why other people find it a struggle, but they are the exception! Most people find difficulties with some concepts at some stage, and need to make an extra effort to get back in the swim.

A very experienced mathematics teacher gives this advice to pupils:

'If you feel that at any time you are falling behind, it is vital to ask for help and support immediately because it can be difficult to recover if you leave the problem unresolved for too long.'

Some people find mathematics and physics harder than chemistry. Others find chemistry much more difficult than biology. Often people who enjoy biology and do very well at it are not confident in mathematics. Others prefer a topic-based approach, such as environmental science, where they find they can understand the concepts more easily because they are applied to a specific area or problem.

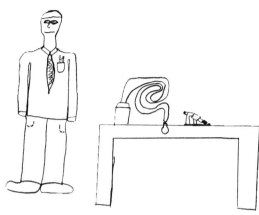

Scientist at work by James

Science concepts are not just 'common sense'. Most people find that the concepts have to be worked at before they feel confident in discussing scientific ideas. Some scientists believe that the difficulty of the basic concepts puts many people off science before they ever reach the more rewarding and creative aspects of the subject.

Is science boring?

Science is like learning to play a musical instrument: the longer you work at it, the more satisfaction it will give you and the more creative you can be. If science at school is taught well, it should convey the excitement of having ideas and testing them out. A dull teacher can make science seem like a list of facts and ideas that are either right or wrong. In that kind of class there doesn't seem to be much room for debate or self-expression. The school curriculum is being revised to make science lessons more relevant to everyday life and current issues.

Because science is a fast-changing field, and every new discovery leads to new theories and new concepts, the science being taught on school and degree courses is continually developing. You may have learned about developments in genetics and astronomy at school that people who were at school or college when your parents were young had never heard of.

Alongside scientific changes, there have been huge developments in information technology. Computers have enabled scientists to develop highly sophisticated instruments, to store and analyse huge quantities of data and to make scientific models and simulations. They can use scientific databases on the other side of the world and communicate with each other through email and the Internet. The huge development of computer graphics helps scientists to communicate ideas, not only within the scientific community but also outside it.

Is science narrow?

When you leave your degree course you will know more about your subject, and have greater computer skills, than people who left a similar course ten years ago. You will have marketable up-to-date knowledge and technical ability, as well as all the other qualities that employers expect of graduates, such as skills of analysis, researching and problem-solving.

The employment situation is very different for someone with a degree in English literature or history. There are very few jobs where the content of these courses is immediately useful; arts graduates have to rely on other more general skills or take vocational training courses. Science graduates can do almost all the jobs that arts graduates go into - particularly if they have good communication skills - with the added advantage of being able to use their scientific knowledge, either directly or indirectly.

Science is international

Another thing that is different about science is its internationalism. Scientific ideas and applications are not limited by national boundaries, or by local custom and practice. You will see in the career profiles featured in this book that science graduates can find job opportunities overseas, as well as opportunities to study and collaborate with scientists from other countries. There are chances to study abroad through exchange schemes at school, and by choosing a higher education course which includes a period overseas, such as with the Socrates/Erasmus scheme sponsored by the European Union. Through information and communications technology, scientists are aware of the work of others and are in constant contact with colleagues around the world. This global outlook is a feature of the science departments at most universities and will be a useful asset even if you go into non-scientific work after your higher education.

The international nature of science means that there are many opportunities for science graduates to work or study overseas at some stage of their careers. International organisations, world trade and commerce also create opportunities to travel and work abroad. One of the aims of the European Commission is to enhance the mobility of scientists between member states.

Science opens doors

Taking a science degree is clearly not just a route into being a highly qualified lab assistant. There are many ways to become involved in science without doing practical laboratory work. On the other hand, many scientists complain that they don't get enough time in their laboratories!

A science degree is a starting point for a lot of different career routes. It is a general education which will help you to understand some of the most crucial issues affecting the future of this planet.

There has been a huge increase in popular interest and debate about scientific developments, particularly in medical and biological research and in computer applications. For example, the possibilities of 'designer babies', of electronic tagging, or of transplants using tissue from human foetuses or from other species all raise ethical issues which should, and do, lead to public discussion. There is an increasing demand and need for good communicators who understand scientific developments and their likely impact. This need, together with the information technology explosion, has opened up exciting new career developments in and around science. Science is definitely not for the inarticulate.

Science means more career choice

Most of the information about careers for science graduates divides the opportunities into three groups:

1) careers using specialist scientific knowledge
2) careers using general scientific knowledge and other qualities or skills
3) careers using the other aspects of your science education (ICT, numeracy, logic, etc) and other qualities or skills.

Most science graduates have jobs from different categories at different stages of their careers. But again the 'one-way escalator' effect means that most people move in the direction of 1 to 3, and many start at 2 or 3.

Rebecca Dumbrell

Rebecca Dumbrell is an example of somebody who used her scientific knowledge to move into a related career in sales.

Career Profile

Age: 24

Technical sales representative

A levels: chemistry, biology, geology

BSc: Biochemistry at University of Bath

'After passing all my GCSEs, I wanted to choose A levels which would give me the chance of employment in future life. At the time, the papers were all saying that, because of the demand for more medical research, in a few years the world would be crying out for biochemists.

I therefore took physics, chemistry, biology and geology A levels at Bexhill College in East Sussex. I had always enjoyed science, but dropped physics as I found it to be too mathematical. However, geology was counted as a science, so I still had three to go to university. Geology was also very chemical which I understood and enjoyed.

I passed biology, chemistry and geology and applied to several universities to study biochemistry. After visiting several, I fell in love with Bath which also had an excellent reputation for biochemistry (one of the top in the country, I believe) and I was accepted.

The course lasted four years, with two six-month placements in laboratories. The second of my placements was in Gottingen University in Germany doing molecular genetics - absolutely brilliant. I went there speaking no German, and I came back thinking in the language! The project there was fantastic and put the whole biochemistry course into perspective - it was totally different from the course practicals, which everyone hated.

I wanted to return to Germany after finishing my course to do a PhD, but lack of finances forced me into going for a job in the UK. An employment agency directed me towards Elhay Lab Products, based in Bath itself and offering a job in sales of plastic consumables. I would be selling a quality product to scientists. Although it was not what I originally set out to do, I have been working for the company throughout the south and west for 18 months now, and I very much enjoy it. My customers are lovely and every day is different, so I am never bored.'

Only you can choose

This book should encourage you to think again about some of the stereotypes we all carry around. Information about careers and jobs comes to you from many different sources, and you have been absorbing it ever since you were very young. Now you are able to be more discriminating and can look more critically at the information you pick up, whether formally or informally. You are able to question and assess information and opinion, to judge how relevant it is now or how useful it will be to your situation in the future. Certainly, if you are going to be a scientist you will need to assess data critically. Ask yourself how the information was collected, by whom and to what end?

Things change over time. Are the facts you obtain now likely to be true in four or five years' time when you finish your degree course? Things which look attractive now may not seem so interesting after a few years of study. New opportunities may arise which do not yet exist. New influences, both economic and social, could affect the prospects for scientists by the time you are in your twenties or thirties.

You will change too. You will develop new skills and interests. You may find you are much better than you thought at some things which you now find difficult. You could find you are brilliant at something you have not even tried yet!

Your values will change. You might find, for example, that you become passionately involved in some environmental issue and need a knowledge of chemistry to understand the problem or contribute to a solution. You might be attracted to a particular career now because you have heard the starting salaries are good. But what happens after the first few years? Will you continue to be interested in the job? Will it seem worthwhile? Will you be able to progress and will your skills be useful in other areas later on?

There are no easy answers to any of these questions. But it's no use doing nothing and sticking your head in the sand, hoping it will all turn out right in the end. It's your career, and you are the best person to have control of it. Take time to look at all the options.

This book will help you to ask the right questions of yourself and other people. It is a starting point for your decisions about science subjects at A level and equivalent and in higher education.

Science in the future

Predictions of scientific breakthroughs in the near future include:

- greater understanding of how the universe is expanding
- nanoscience - the development of products the size of a nanometre (a millionth of a millimetre) - will result in even further miniaturisation of electrical circuits
- advances in the use of computer models to predict climate change more accurately
- the mapping of the human genome, due to be completed in the next few years, will identify every gene and its function
- fusion energy will offer a truly clean, safe alternative to carbon-based fuels.

But, Sir Harry Kroto, Nobel prizewinner in 1996, has pointed out that important discoveries in science are usually totally unpredictable. That's what makes them important!

Science graduates tell their stories

Throughout this book you will find profiles of science graduates in their twenties and thirties, explaining the subject choices they made and their career paths so far. The profiles are genuine stories of people with degrees or diplomas in science and mathematical subjects. As you read through the book you will find that some of the things they say have also been quoted in the relevant sections. Here are some extracts from some of them:

Garry Hepherd - page 78

(textile chemistry graduate, now an account manager)

Garry came up against the science/arts divide, and *'was forced to make a choice between a science or arts biased subject route at an early age. I found this dogmatic approach to learning particularly limiting since I was not allowed to combine my three favourite subjects - chemistry, history and French at A level'.* Fortunately, Garry's choice of three sciences has worked out well for him.

Jane Davies - page 85

(agricultural botany graduate, now a science teacher)

'*I chose to maximise my career possibilities by selecting scientific A levels*', says Jane, who had always considered teaching as a career but wanted to keep her options open until the postgraduate level.

Rebecca Dumbrell - page 14

(biochemistry graduate, now a technical sales representative)

Rebecca considered the job market when she chose her degree course: '*The papers were all saying that, because of the demand for more medical research, in a few years the world would be crying out for biochemists.*'

Jessica Harris - page 120

(mathematical statistics graduate, now a medical statistician)

Jessica's career has shown how other skills, such as communication, can be combined with your degree subject knowledge to add depth to a job. As well as designing, carrying out and analysing surveys, Jessica is '*also responsible for contributing towards reporting results in medical journals ... and at scientific meetings*'.

Sue Wood - page 44

(biological sciences graduate, now a scientific officer)

Sue describes how a science degree course is anything but boring. '*Highlights of the degree included lots of laboratory and fieldwork experience, well organised up-to-date courses, an ecology field trip in Yorkshire and sponsorship to attend the first global student conference on the environment in Turkey.*'

Mike Partridge - page 118

(natural sciences graduate, now a research fellow for the Institute of Cancer Research)

After a spell in administration, Mike decided that he really wanted to return to '*being a scientist*'. His current work is '*academically stimulating and personally very rewarding*'. After several career moves, learning new specialisms, Mike found that '*the core skills of a research physicist remain the same, and skills learned in one field were easily transferred to another*'.

Paul Compton - page 109

(ophthalmic optics graduate, now working as an optometrist)

Paul can vouch for the value of work experience placements while still at school. His interest in medical science and research became focused after a period with a local optometrist, which '*gave me valuable insight into a career which previously I had known very little about. I applied to the five universities offering degrees in ophthalmic optics ...*'

Rob Brown - page 146

(biotechnology graduate, now a molecular biologist with a pharmaceutical company)

Rob's career also grew out of his work experience placement, but at undergraduate level rather than at school: '*the experience and links I made ... were sufficient to secure me a job and the opportunity to study towards a PhD at the same establishment where I graduated*'.

Celia Keen - page 139

(chemistry graduate, now a Chartered European Patent Attorney)

Celia wanted a broader career than that of a research chemist. '*I wanted to use my chemistry but to embrace other areas of expertise too. The career of a patent agent combines science with law and also involves a linguistic component, so this seemed the ideal choice*'.

Paul Pilkington - page 89

(physics graduate, now a graduate trainee with British Aerospace)

Paul studied physics although he always intended to go into engineering, because: '*At school ... I was always taking things apart and putting them back together again.*'

Chapter 2: The image of science and scientists

This chapter covers:
- ☐ views of science and scientists in Britain
- ☐ some comparisons with other countries
- ☐ how subject choices are influenced in schools
- ☐ advanced level subject and degree subject choices
- ☐ changes ahead.

Just imagine

When you picture a scientist, do you think of a man or woman who is:

old	young
intelligent	stupid
hardworking	lazy
scruffy	smart
kind	mean
dull	fun
rich	poor
cold	friendly
outgoing	shy
a back-room person	a leader
creative	plodding
tone-deaf	musical

We are all influenced by stereotypes, and not just when it comes to scientists. Science seems such an unknown quantity to many people that, as a result, some very strong stereotypes have emerged to depict the people who study and work in the area. Even Celia Keen, a highly qualified scientist herself (see page 139), talks about 'the archetypal garden-shed inventor'! The two most common caricatures are the eccentric boffin - brought to life on TV by personalities like David Bellamy, Adam Hart-Davis, Heinz Woolf and Patrick Moore - and the evil professor intent on destroying his enemies and ruling the world.

The eccentric boffin is obviously brainy, won't win the best-dressed man award (it's always a man; Carol Vorderman doesn't qualify!), and has an obsessive interest in strange and obscure things. He gets very excited, shouts a lot and waves his arms about. The evil professor is not nearly as lovable. He is a scientist who has gone off the rails and is power mad. He threatens the planet in books and comics, on TV, films and computer games. There are other stereotypes like the grey men in thick glasses who work on incomprehensible problems in locked rooms, and bizarre mutants who have been affected by experiments that have gone out of control.

Stereotypes like these are extreme examples that show the fear and fascination that science brings out in people. If you did a survey you might find scientists who are a bit like all of them, but you would also discover that the kinds of people who study and work in science are as varied as doctors, lawyers, teachers or any other professional people.

Think carefully about these images and how they could affect your interest in science.

Does studying science mean becoming a scientist?

In chapter 1, the point was made that people seem to confuse science the subject with science the career in a way that does not happen with other subjects.

This observation was stated more strongly by a BBC science editor at a conference of the British Association for the Advancement of Science (BAAS) a few years ago:

'The problem starts at school. We do not teach students history or English and assume they will become historians or novelists; we do not teach French or German to equip students to become simultaneous translators; so why is science taught on the basis that students are going to become practising scientists?'

What do you think? Should science be part of everyone's general education, or is it something that is only important for people who want to become scientists?

Chapter 2 - The image of science and scientists

Science is for all of us

Today, everyone must study some science at school until the age of 16, but this has not always been so. International surveys show that the basic level of scientific knowledge among the general population is very poor in Britain when compared with our industrial competitors like Japan and Germany. There is a 'crisis of science' as not enough students are choosing physical science and engineering post-16 and higher education courses.

Does this matter? Yes it does! COPUS, the Campaign for the Public Understanding of Science, is trying to promote science awareness for all, and is trying to break down the divide between 'science people' and 'arts people'. Another organisation, called Save British Science, is pressing for government investment in science and engineering research to double to match that of some of this country's competitors - including paying research scientists more.

Science and technology affect nearly all aspects of daily life, both at home and at work. As this dependence grows, a broad awareness of science and technology issues is essential for informed decision making.

Are we anti-science in Britain?

There is widespread public anxiety about issues such as the uses and abuses of genetic manipulation, pollution, global warming, world energy resources and the possibilities of nuclear terrorism. In Britain, science is often cast as the culprit, with scientists no longer trusted to find solutions. Many people today are suspicious of science and feel that technological progress is a threat to the environment. The exception is in medicine, however, where science is usually looked upon as our saviour (as long as genetics and cruelty to animals are not involved!).

Meanwhile, we all happily go on buying the latest electronic gadgetry and wondering at the pictures sent back from outer space!

Britain has a distinguished history in science and should have a great future. British scientists have collected numerous Nobel prizes. The scientists in our research laboratories and universities today are capable of following in the footsteps of their famous predecessors like Sir Isaac Newton and Charles Darwin - although commercial pressures may restrict 'pure' research to some extent.

Do scientists communicate well?

Some people feel that scientists have been partly to blame for their own bad press and poor public relations because of the kind of language they use and because the scientific community is seen as elitist.

Improving the communication skills of those who 'do' science is fundamental to making science more accessible to the public, but it is not just a one-way problem. All too often, non-scientists are quite happy to say - in fact even to boast - that 'Science is just beyond me', or 'I couldn't understand a word they said', without making any effort to learn.

We think we can understand an artist, writer or composer because we can respond to their work and form an opinion about a painting or piece of music. With science it is not so easy, because the work often cannot be described in readily accessible terms. Even a simplified explanation seems incomprehensible to many people. A certain level of specialist vocabulary and understanding of concepts is needed, and this cannot be picked up in ten minutes. Scientists even have this sort of communication problem when presenting their results to colleagues working in different fields.

Because of this difficulty in communicating, much of the excitement of scientific research and the creativity of developing ideas is lost to people outside science. Scientists may demonstrate their work with great zeal, but they are often trying to share the beauty of an idea or concept with a rather baffled audience.

Things are changing as public interest in science grows and more scientists learn to put across their knowledge and enthusiasm. Books like Stephen Hawking's *A Brief History of Time* and Dava Sobel's *Longitude* have been unexpected bestsellers in recent years.

Scientists in the media

If you want to, you can hear a great number of scientists talking about science. Many young people listen to science programmes on the radio and watch TV programmes like *Horizon* and *Tomorrow's World*. Many also read *New Scientist* and follow science articles in the press. Good communicators and popularisers of science, such as Richard Dawkin and Steve Jones (the genetecists), Heather Couper (the astronomer), Ian Stewart (the mathematician) and Susan Greenfield (who talks inspiringly about how the brain functions), have gone some way to correct the stereotypes through science programmes on TV and radio. However, there is hardly any opportunity in the media to get to know scientists as personalities in the same way as sports stars, politicians, actors, musicians and writers. Think about TV panel games, political discussions like *Question Time*, and chat shows: how many scientists appear on these programmes? A tiny minority, like Patrick Moore, have become household names, but, in the main, science and scientists are kept in a world apart, not wholly of their own making.

Social and moral issues

There are many moral, social and political questions raised by developments in science. Should animals be used to test drugs, for example? Should research into weapons of mass destruction be banned? Should food be irradiated to preserve it for longer? Should organs from specially reared pigs be used for xenotransplantation into humans?

These aspects of science are easier to understand and to voice opinions about. They lead to more discussion and seem, to most people, to be much more interesting than the painstaking work that leads to such questions being posed. It is very difficult to participate in discussions about the technical side of scientific developments unless you are very well up in the subject.

We want the answers

Another problem that seems to baffle non-scientists is why scientists don't know all the answers. School science often seems to deal with questions that are either right or wrong. There are facts to be learned and simple theories which can be tested by a laboratory experiment. It surely follows that there must be right and wrong answers to all scientific questions. Is the greenhouse effect getting worse? What is a safe dose of radiation? Is there a link between cancer and diet? Where and how did AIDS begin?

Scientists are heard giving different and contradictory opinions about topics such as these, and people say, 'Why can't they get the information and give us the facts?' The trouble is that the 'facts' are often far from clear. Somehow the feeling is that scientists have failed if they can't give a straight answer, but there is no simple solution to many of these complex issues. The scientists are giving opinions based on their interpretation of the data available, and good data is often very difficult to obtain.

How do we learn about science?

Controversial aspects of science are very newsworthy and often high on the political agenda. You and your friends are getting a lot of different messages about science and scientists from inside and outside school. This obviously has an effect on your subject preferences and career choices. Research on career choice shows that the stereotypes we discussed earlier are adopted very early on in primary school.

At primary school

A recent survey, asking over one thousand primary age children in England to draw a picture of a scientist, resulted in both boys and girls over the age of six drawing predominantly male scientists. Girls younger than six drew predominantly female figures, indicating that children do have open minds and tend to project themselves into situations before the stereotypes take hold. Baldness, beards, spectacles and white coats featured strongly. Backgrounds were almost universally indoor scenes, and often featured flasks, test tubes and beakers. Only 15% showed living things. Boys were more likely than girls to show backgrounds connected with energy and space science.

Most primary school age children will have little or no personal knowledge of scientists, so these stereotypes are presumably drawn from books, comics, television and videos.

We commissioned our own small and unscientific survey of nine-year-olds' perceptions at Aloeric School in Melksham in the spring of 1999. An admirable percentage of these children did draw female scientists. The teacher assures us that the children were not briefed to do so; they just happened to be a group who were particularly hot on equal opportunities. Several of them included computers in the picture. One seems to show an Antarctic explorer. We have used some of their drawings to decorate the text in this book.

We also asked the children to write a few words about being a scientist. Here are some of their impressions - with the original spelling:

'I would like to be a scientist because you can invent things and use medicil liquids and do experiments. They invent computers, TVs, telephones, cd players and loads of things. I think it would be fun because you work in a lab.'

Scientist at work by Matthew

'I would like to be a scientist because I would like to be famous and i would like to invent things and make money to so I could be rich and be in a scientist book.'

'I would not like to be a scientist because if you mix dangerous chemicals wrong and it might blow up.'

'No because you have to get up in the morning about 5 AM and if you don't make a discovery you don't get paid a lot.'

'I would like to be a scientist because you get to travel to Antarctica you get to invent things Make chemicals, have a lab, find out new metals and look for oil.'

'I would not like to be a scientist because you have to spend all your time thinking and inventing and it takes a long time and you don't get to go out and once you have done one invention you have to think of another one and another one and you can't see your friend because you're to busy.'

'I would like to be a scientist because I do like to explore and would not mind working for 12 hours a day.'

'No way! I would not like to be a scientist because me I'm just not the right person I'd probably never get any money because I wouldn't be able to do anything apart from blow up the place and I'd never discover anything anyway.'

'I would like to be a scientist because I would invent a yoyo that sleeps for a week so we can do every trick in the world and a car that works on its own.'

'I don't wont to be a scientist because I wont to be a Football Player.'

How do older schoolchildren see scientists?

An on-going international survey shows that perceptions of scientists among 13-year-olds show a distinct difference between the industrialised countries and the developing world. In the latter, children see science as interesting and exciting and scientists as people who will help to grow successful crops and cure the sick. In the western world, we are more suspicious and questioning.

The images around you of science and scientists will have some effect on your choice of career and the subjects you decide to study after GCSE. But your opinion of science will also be influenced by your experience at school: one of the most important factors for most people choosing science is an enthusiastic and interesting teacher.

If you are at the 'options' stage of your education, you might like to reflect on these words from the Royal Society's leaflet *Thinking about Science?*:

'The Royal Society strongly advocates double-award science courses for all pupils, including those who will become professional scientists. Such courses give breadth and depth of study, encourage practical, investigative science and can provide clear motivation for pupils to continue with science. They give a firm foundation for A level sciences or Science GNVQ at Advanced Level. Pupils who choose other subjects beyond 16, or who leave formal education, will have a broad and valuable understanding of science which will prepare them for life in a scientific and technologically based society.'

They also say that a single-award science 'seriously limits choices later' and that the 30% of curriculum time taken up by all three separate science GCSEs (chemistry, physics and biology) may be 'too high for a balanced curriculum' unless you are already pretty certain that you want a scientific career.

Advanced level choice

It is at this level where the effects of subject preferences really begin to show. This is when you must narrow your subject choices to two, three or sometimes four subjects in the present A level system. This changes in September 2000, when the new Advanced Subsidiary courses, which allow a broader base of up to five subjects in year 12, bring the English and Welsh system closer to that of Scotland, where the choice is four or five subjects for Highers or Advanced Highers. This gives you more chance of keeping your options open by mixing sciences with other subjects.

Advanced GNVQ in science, IT, engineering or manufacturing can be taken as the equivalent of two A levels, although most higher education institutions will require a science A level to be taken as well for entry to a degree course. Single award Advanced GNVQ (six units, equivalent to one A level) is being piloted and should be in place by September 2000, as well as a three-unit Part GNVQ from 2001.

An alternative route into employment or higher education for school-leavers is to take a Modern Apprenticeship with a science-based or engineering company, which will take you to at least National Vocational Qualification level 3.

Reasons for choosing certain subjects at Advanced level include:
- getting good results at GCSE
- finding the subject interesting and enjoyable
- being encouraged by teachers who think you can pass at Advanced level
- needing that particular subject for entry to a higher education course or career.

Enjoyment of a subject and good results at GCSE are the main factors for most people, but for those choosing science subjects, particularly physics and chemistry, the fact that they need the subjects for a course or career is more likely to be important. So the choice of science post-16 is affected by both subject preferences and career choice.

Are fewer people choosing sciences?

In recent years, when A level results have been published, there have been headlines in the newspapers about fewer and fewer people taking A level physics. In fact, the number of entries for all the sciences have increased in the last couple of years.

Careers with a Science Degree

A level entries in selected subjects 1990-98

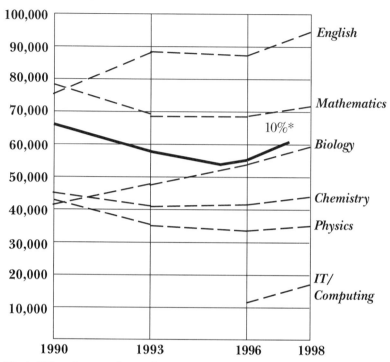

*The bold line shown on the graph represents ten per cent of the age group (excluding Scotland) of A level age for each year, as a comparison.

Key points

- The number of 18 year-olds has been increasing over the last few years, following a period of decline.
- While the overall proportion of 18 year-olds sitting A level subjects rises each year, these 'extra' students are not choosing to study maths and physics. They are more likely to be choosing social science and humanities subjects such as English, history, economics and sociology.

What is the chance of getting good grades?

The provisional data based on 1998 A level results in maths, sciences and IT show the percentage achieving each grade as follows:

Subject	A	B	C	D	E	F/N	U
Maths	28%	19%	17%	14%	11%	7%	5%
Physics	22%	21%	19%	15%	11%	7%	4%
Chemistry	23%	21%	19%	15%	11%	7%	4%
Biology	16%	19%	19%	17%	14%	8%	6%
Computing/IT	8%	13%	20%	21%	18%	11%	9%
All subjects	17%	19%	21%	18%	13%	7%	5%

Source: Qualifications and Curriculum Authority

(figures may not add up to 100% because of being rounded up or down)

Key points

☐ Results vary from year to year, and have tended to improve in recent years. In 1994, for example, only 18 per cent achieved grade A in physics and chemistry, and 8% and 10% respectively were unclassified.

☐ The proportion of students obtaining A or B grades in maths, physics and chemistry was higher than the average for all subjects.

☐ These figures do not include those who gave up the course and did not attempt the exam.

What happens in higher education?

Students choosing degree courses

The figures for those starting degree courses in the main science subjects in 1994, 1996 and 1998 are given below. The proportions of all students choosing the subject are shown by the percentages.

Students accepted onto degree courses in sciences

	1994 No.	%	1996 No.	%	1998 No.	%
Biological sciences	14,877	6.0	16,578	6.1	17,067	5.6
Physical sciences	14,608	5.9	14,704	5.4	14,729	4.9
Physics	2,930	1.2	2,902	1.1	2,997	1.0
Mathematical sciences/informatics	14,050	5.7	17,724	6.5	21,717	7.2
Engineering and technology	23,756	9.6	22,891	8.4	22,016	7.3
Combined sciences	5,516	2.2	5,980	2.2	6,717	2.2
All subjects	247,567		273,032		302,683	

Key points

- The percentage of those applying who choose science subjects is decreasing, although numbers are increasing. This is because of the rapid expansion of student numbers in higher education, most of which has been seen in the arts and humanities.

- The only increase in the subjects shown is in mathematical sciences/informatics, presumably down to the information revolution and the growth in employment in financial services. Combined science entries have remained steady.

- More students going to university recently have been over 20 years old - although numbers have fallen since the introduction of tuition fees. These older students tend to be more attracted to subjects like engineering, social sciences and business studies rather than pure sciences and mathematics.

- Medicine and veterinary science (not shown in the table) continue to attract applicants with strong science qualifications. Entry to these courses is limited by the needs of the profession, and many of the rejected students find their way onto other science-related courses. More places in medical schools are planned, which may affect demand for other courses.

- The numbers doing Higher National Diploma courses in sciences are very small in comparison with the numbers on degree courses.

Getting into higher education

The ratio of applicants to places in higher education is not straightforward to calculate because each applicant can choose up to six different courses. In general, there are many fewer applicants per place for science courses than for arts, social science or business studies. However, this does not necessarily mean that the quality of the applicants is lower. Maths and physics, for example, do not attract as many applicants as other subjects, but those who do apply tend to have high A level grades. Medical and veterinary courses are known to be very competitive, so most of the applicants are expected to achieve high grades. There is more information in chapter 4 about the requirements for courses in higher education.

How do students view themselves?

Past surveys have shown that there are more similarities between students than there are obvious differences. Contrary to some expectations, you will be able to find history students who say that they have technical minds and dislike working with people. You will equally find science students who describe themselves as imaginative and having a broad outlook on life. And people who are good at a subject tend not to find their courses difficult. So pushing yourself into subject choices by trying to match yourself to the perceived qualities of a typical student is not likely to be productive. It is much better to follow your own abilities and interests.

What about the careers aspects?

First employment figures for graduates get a lot of publicity. Lots of anecdotal information feeds back into the schools and colleges where you and your friends are making subject choices and university applications. Some of the stories will be encouraging; others may put you off.

Science graduates who opt to do a PhD as a three-year full-time higher degree by research, leading to the title Doctor of Philosophy (see chapter 7 for more details) face another three years on a very low income. However most feel that it is well worth it. Salaries are likely to be £2000 more for a PhD holder than for an average first degree graduate. Science graduates sometimes feel that their pay levels do not reflect the contribution they make. The ones who go into careers such as finance and management after graduating often enjoy higher starting salaries than scientists in industry or in the academic world.

A science degree does not guarantee a job, but it can give you a clear advantage by opening up a broader range of careers than an arts degree. In 1998, 21% of graduate recruiters reported a shortage of scientific/technical/engineering/research and development staff, compared to 16.7% in 1997. There will be an increasing need in the future for science and technology graduates in a wide variety of jobs.

Degree subjects like accountancy or law may sound much safer, but science will give you more flexibility in your career plans. As you will see later in this book, science graduates can leave their options open and choose later from areas of work both in science-related and non-scientific disciplines. The important thing is to gather as much information as you can before making your decision.

Science awareness is growing

The Government is aware of the shortage of science and technology graduates, and this is reflected in the amount of funding that these faculties receive.

Some higher education institutions offer foundation courses or first years of a degree course which are especially designed for people who have taken arts and humanities subjects post-16 - although they will still be expected to have good results in science and maths at GCSE level.

There is now more science in primary schools; science has been added to the list of GCSEs that prospective primary teachers must have at grades A-C. Some primary and most secondary schools are involved in practical science and technology projects which are supported by employers and industrial organisations.

There are also many science fairs and exhibitions welcoming children and adults. Activities like these and centres like the Exploratory in Bristol are much more 'hands on' than science museums used to be. They involve visitors of all ages in the experience of science. There are other science lectures and events run by organisations such as the Royal Society and the separate science institutions and societies. Ask your science teacher for more information.

Other initiatives are aimed at science teachers rather than students, and there are schemes where teachers can spend time in industry working with local companies to develop projects for their schools. COPUS, as explained earlier, aims to raise scientific awareness among the general population.

Science in other countries

Science has a superior status in many other countries. Its lower status in this country is sometimes blamed on the traditional English public school bias toward the classics and humanities. Even families which made their fortunes through advances in technology in the industrial revolution wanted their offspring to have this sort of classical education.

In France, Japan and Germany the level of scientific knowledge in the general population is thought to be much higher than in the UK and the USA. America, like Britain, compares itself with its industrial competitors and is concerned about the general standard of science education. On both sides of the pond, large companies are supporting school science programmes because of worries about the low level of scientific understanding in their workforce.

A new image

The initiatives mentioned above and the concern that drives them should lead to changes in the perceptions of science and scientists. With all young people now studying some science at school, the standard of scientific knowledge is bound to improve. This should mean that they will be more willing to question the stereotypes and to have more positive feelings about science and scientists, although they may still see science as a difficult subject.

Science graduates are increasingly going into other work such as accountancy, where graduates are traditionally better represented in top management. As a result, in time, there will be more science graduates in top jobs in industry and government. There is a specialist 'fast track' scheme for science and engineering graduates in the Civil Service - or they can choose the generalist scheme instead.

The science faculties in universities are becoming more aware of the importance of communication skills and are offering special training. There is even a degree course entitled Media Science, which should appeal to students interested in journalism, PR and marketing as well as science. Higher education institutions now realise the value placed on high standards of literacy and numeracy by employers.

Links with the rest of Europe, where science has greater prestige, are being strengthened. Programmes, such as Socrates- Erasmus and Leonardo, foster the exchange of scientific personnel between European countries. Increased contact between governments and through other European agencies should help to increase the status of British scientists.

The number of women in science and science-related jobs is gradually increasing so there are more role models for younger women. More women scientists and science teachers will help both sexes to deal with some of the damaging stereotypes. Awareness of equal opportunities issues in science is improving at school and in science careers.

Women have always been well represented in communications and there are many more opportunities now in science publishing, journalism and the other media. Again, these women are in a position to have a good influence and act as role models for young women beginning their careers.

Summary

So, there are many good reasons for examining the images of science and scientists more closely, rather than allowing the stereotypes to limit your choices and your career aspirations. Things are changing and will continue to change.

- ☐ Developments in science lead to advances in technology. Understanding these developments will enable you to deal with the changes that are coming and their consequences. A scientific education will allow you to take a part in shaping the future.

Chapter 2 - The image of science and scientists

- ☐ There is a shortage of science, engineering and technology graduates, perhaps because of a lack of awareness of the range of jobs open to science graduates.

- ☐ There seems to be a general assumption, in Britain at least, that if you are interested in science you will become a research scientist. This misconception appears very early on in primary school.

- ☐ Despite general public interest in issues raised by developments in science, scientific understanding is still not widely seen as an important part of our culture. Efforts are being made to interest the general public and increase science awareness in many different ways.

- ☐ The influence and prominence of scientists will change as more young science graduates move into a wider range of careers.

- ☐ Stereotypes of scientists should disappear, as the number of science graduates in the general population increases and scientific knowledge improves generally.

Careers with a Science Degree

Chapter 3 - Women and men in science

This chapter covers:
- ☐ sexual stereotypes of science and scientists
- ☐ male and female subject choices at A level or the equivalent, degree and postgraduate level
- ☐ influences on girls and boys making subject choices
- ☐ women at work/discrimination
- ☐ how girls can get support when choosing science subjects and scientific careers.

In chapter 2 one factor that kept recurring was gender - the male stereotype of a scientist. This chapter considers the problem in more detail and asks why many girls are turned away from science. It also looks at the data on female and male subject choices at each stage of education, and speculates why the differences arise. This is a topic everyone has views on, so be prepared to argue and disagree. It is important to understand the influence of gender: how it affects the way you think about science and the decisions you make about your subjects and career.

But first... a test

Spot the female scientist!

Hint: there is one woman on each horizontal row.

I Newton	C Darwin	D Hodgkin
J Watson	L Pauling	B McClintock
M Curie	A Einstein	F Crick
S Hawking	R Franklin	F Sanger
A Fleming	C Herschel	Frankenstein

(answers on page 53)

How did you do? Can you think of any other women scientists? It's not surprising if you find it difficult because women scientists are still outnumbered by men today and there were even fewer in the past.

Fact: 85% of the full-time science, engineering and technology workforce are male.

In 1998, more girls than boys achieved higher grades in GCSE combined sciences, although the picture is reversed for individual physics, chemistry and biology GCSEs. At A level, girls accounted for only 22% of the total physics candidates; only 21% of computing candidates were female. Why is this?

What happens in education to influence both girls and boys in career and subject choices? And what has been happening in the past to discourage girls from achieving their potential as scientists?

A woman's place

Historically, women were not expected to play an active role in science, and even when they did they were not given credit for their work. For example, science historians have recently discovered that one of the most widely-read early medical texts in Europe was written by an Italian woman called Trotula. She was one of a group of medical women known as the 'ladies of Salerno'. Over the years her name had been left off the text and it came to be assumed that it was written by a man. In more recent times many people were outraged that Rosalind Franklin was not properly recognised for her contribution to the understanding of DNA.

But some women have got recognition. Marie Curie won two Nobel prizes and later her daughter Irene won one for physics. More recently, when Barbara McClintock suggested that genes could 'jump around' on a chromosome, the members of the American National Academy of Sciences were less than enthusiastic. She says, 'They thought I was crazy, absolutely mad!' But she was right and won a Nobel prize in 1983.

A man's field?

You have probably noticed that one of the scientists in the test was very different from the others. Dr Frankenstein is the scientist in

Mary Shelley's famous story. His drive to conquer nature by giving life to his creature, with no thought for ethical or humane considerations, still feeds many common prejudices about science and scientists.

The pursuit of science is often portrayed as competitive rather than collaborative, with the outcome being to control nature rather than to understand and work with the natural world. The language used in the media to portray science often has a 'masculine' sound to it - conquering, controlling, probing to the depths, overcoming, a race against nature, unlocking the secrets of mother nature, penetrating the secrets of the universe!

Studies of the scientific staff working on the American Space Programme at NASA showed that while many were indeed cool and logical, some of the most creative researchers were found to be emotional, intuitive and impulsive. Many of the most innovative

> 'Somehow lots of women in Britain have lost out on the enjoyment of science.'
>
> *Dr Elizabeth Johnson, an American physicist working in Britain*

research scientists, both men and women, show these traits. Perhaps these people are less inhibited about asking unusual questions and are able to make conceptual jumps more easily. Scientific research, like most other areas of work, needs a range of different people contributing different things to the job.

But not all science graduates become research scientists, and there are many other opportunities for people qualified in science. These jobs require all sorts of other qualities and skills and offer rewarding prospects, but many girls do not get to the starting post because they have already turned away from science.

Boys too will have been influenced by the kind of images and stereotypes discussed in chapter 2. Perhaps some are choosing science because it seemed expected of them, when they might have preferred something different.

So where do these perceptions of science and scientists start?

Back to primary school

Throughout the book, there are pictures of scientists drawn by a class of nine-year-old children.

As already described in chapter 2, a survey to determine children's perceptions at the time of the introduction of science in the National Curriculum found that children have acquired some stereotypical images by the age of six - the male scientist who is bearded, balding and wears spectacles and a white coat. If these are the images which children of that age have, the influence of books, comics and the television must be enormous.

Is it any wonder that the girls particularly find it hard to imagine themselves in the role?

Secondary school

At GCSE, boys' and girls' subject choices are, nowadays, drawing closer together because the National Curriculum requires everyone to do a science subject. The proportions of boys and girls getting higher grade GCSE passes in sciences and maths are very similar.

In English language, however, the gap is much wider, with boys' results at GCSE way behind the girls'. This gets much less publicity, but it may also have a bearing on boys' subject choices. Some may opt for science because they are less confident in the language skills needed for subjects that require more essay writing.

Studies show that both boys and girls seem to make up their minds about studying science early on in secondary school. Unfortunately, girls are more easily put off. If they find the subject becoming difficult they are more likely to conclude that it is too hard for them and lose interest. Boys who have difficulty with science and maths tend to see themselves as lazy, whereas girls who have problems see themselves as stupid.

The DTI/Promoting SET for Women 'Go For It' poster campaign was launched in November 1998 after research among 14-16 year old girls revealed that:

☐ girls are often alienated by the impersonal and value-free aspect of science

- very few girls had ever met a female engineer
- girls want to work in a socially supportive environment, and don't see scientific jobs as fulfilling that need.

A lot of girls have a suitable science base at GCSE but few of them choose to study the subjects beyond GCSE. Many of these girls have better results than some of the boys who do continue with sciences.

Fact: Over twice as many boys take maths and science-based GNVQs.

Fact: In 1996, out of 5,900 people who started Modern Apprenticeships in Engineering Construction, only 192 were girls.

It may be the way science is taught in schools that puts girls off. One view is that science experiments are too abstract: they are not set in a context which relates to real life and real problems, and so fail to attract the interest of many girls.

> 'Science has in the past been taught as too abstract a subject, and in a way which has ignored values. The whole development within science and technology needs a values component. The field needs the qualities young women can bring.'
>
> *Jan Harding, chemistry education specialist, consultant and writer on gender and science*

What happens in your school? Do you think boys and girls have different attitudes to science, maths and computing?

There are organisations working to encourage more girls to take Science, Engineering and Technology (SET) subjects beyond GCSE, and pursue SET careers. Examples of these include WISE - the Engineering Council's Women into Science and Engineering Campaign, WES - the Women's Engineering Society and Women in Medicine.

Peer group pressure

Studies have shown that boys in mixed schools and girls in single sex schools are more likely to choose science at A level than girls in mixed schools and boys in single sex schools. So peer group pressure from the opposite sex seems to influence both boys and girls in making their choices.

Some schools separate boys and girls for science classes and have 'girls only' sessions in the computer room. Perhaps 'boys only' English classes would improve their language skills and increase their confidence as communicators.

A visit to any school's computer room in the lunch hour will confirm the gender gap. The content of computer games does not appeal to girls as much as it does to boys. This alienates many girls and undermines their confidence in the use of computers for more serious purposes.

What happens at A level?

Because science A levels are required for entry to most science degree courses (see chapter 4 for details), the choices you make at this stage are crucial.

Entries to A level subjects according to gender have stayed fairly constant over the past few years. Figures for 1998 are as follows:

	Women	Men
Mathematics	36%	64%
Physics	22%	78%
Chemistry	46%	54%
Biology	60%	40%
Computing/IT	21%	79%
All subjects	54%	46%

As you can see from the chart, a lot more girls choose biology, which may be required for many of the so-called 'caring' careers in medicine, veterinary science, etc. If you do choose biology it is a good idea to take chemistry too, particularly if you want to study biology in higher education.

It is in physics and computing that girls are so poorly represented, and to a lesser extent in maths.

- Less than two per cent of the total female A level entries in England and Wales were in physics in 1998, compared with seven per cent for males.
- Of the total A level physics candidates, girls accounted for only 22%.

Yet physics is a subject that opens up a huge range of career options in science and engineering. The fact that these subjects are known as the 'hard sciences' may be putting them off, but girls who carried on have shown that they are just as capable as boys.

What about degree courses?

At the next stage after A levels or equivalent courses, the proportions of males and females accepted onto degree courses in 1998 were as follows:

	Women	Men
Mathematical sciences/informatics	23%	77%
Physics	19%	81%
Chemistry	40%	60%
Biological sciences	65%	35%
Computer science	19%	81%
Engineering and technology	15%	85%
All subjects	53%	47%

Source: UCAS

At present, representation of women in undergraduate courses in the biological sciences is good, but in maths, physical sciences, computer science and engineering the representation is significantly less than for men (the proportions of women obtaining degrees in these subjects are 39%, 38%, 20% and 15% respectively).

There are many other degree courses attracting students with science A levels. Women tend to outnumber men on degree courses allied to medicine, veterinary science, environmental science and food science. Men, on the other hand, are in the majority in engineering, forestry and combined sciences.

So there does seem to be a strong female preference for biological sciences, medicine and environmental subjects. As we have observed, society tends to expect different things of men and women and this is reflected in their subject choices. Although these are not the so-called 'hard sciences', they are far from soft options in terms of A level grades and competition for places.

Sue Wood

Sue is a scientific officer who comments, of her future, 'the sky's the limit!'

Career Profile

Age: 22

Scientific Officer at Institute of Virology and Environmental Microbiology, Oxford

A levels: biology, geography, mathematics

Degree: Biological Sciences, Exeter University

Masters degree: Environmental and Ecological Sciences, Lancaster University

'After GCSEs, I wasn't sure what A levels I wanted to do. I was interested in environmental issues and liked science, so I decided on a combination of biology, geography and maths. This combination of subjects would also allow me to go on and study environmental science or biology at university and, as I wasn't sure what I wanted to study, it kept my options open. I stayed at the sixth form at my school as I liked the courses and teachers. Aspects of the A level biology course that I found particularly appealing were an ecology field course and carrying out my individual project. I chose to look at the impact

of human recreational activities on the abundance and distribution of invertebrates at a nearby lake.

I wanted to go on and study ecology and biology at university and my first choice was joint honours biology and geography degree at Exeter University. In choosing university courses, I looked for a course with a high amount of practical work, a broad range of subjects in the first year and the ability to specialise in the last two years. I also wanted to go somewhere where the university and the department had a good reputation.

I didn't quite make the grades required for the joint honours course and I was lucky to be offered single honours biological sciences course at Exeter, which I accepted as I could still take geography modules in the first year. In the second year, and increasingly so in the third year, courses became more specialised. I chose plant biology, ecology and microbiology options in the second year and soil ecology, plant pathology and microbiology courses in the final year. Highlights of the degree included lots of laboratory and fieldwork experience, well organised up-to-date courses, an ecology field trip in Yorkshire and sponsorship to attend the first global student conference on the environment in Turkey.

My involvement with the conference intensified my wish to work in the environmental sector and I busily applied for any appropriate jobs from magazines such as the New Scientist, Environment Post and Countryside Jobs Service. I attended several interviews, which were good practice at improving my interview technique but did not lead to any offer of paid employment. I therefore decided, during the summer after I graduated, to apply for a Master's degree in environmental science with the hope that the extra qualification would give me the specialised knowledge and experience I needed. I applied for several courses, but the Environmental and Ecological Science degree at Lancaster University appealed the most. I liked the high content and wide range of science-based courses, and the substantial amount of assessment was by course work, which I prefer to exams. I chose a wide range of courses including the study of air pollution and climate change, water pollution and supply, ecotoxicology, environmental management, environmental molecular biology and the National Vegetation System Classification field course.

I wanted to gain experience with an outside organisation and applied for an MSc dissertation project with the NERC-run Institute of Freshwater Ecology in Windermere looking at non-obligate bacterial predation. I enjoyed the microbiology courses I had taken at undergraduate level and also the molecular microbiology course at Lancaster so this project was very well suited to me! The change of working environment was very refreshing and I am indebted to both my supervisors for all the help and support they gave me.

I attended another environmental student conference, this time in Stockholm, sponsored by The Graduate College at Lancaster University. The experience was extremely gratifying as I was able to apply my knowledge from my course to the debates we had about how students can campaign more effectively about environment issues, both locally and on a wider scale. I was an environmental officer during the time I was at Lancaster, and in a small way we promoted environmental issues by setting up an organic garden and improving the recycling scheme at the university.

In the summer of my year-long MSc course (from September to September) I applied for many jobs and secured a position at one of the sister institutes of IFE: the Institute of Virology and Environmental Microbiology. I am now working as a Scientific Officer looking at gene expression in bacteria that degrade the pesticide 2,4-D. I like the work I do for several reasons. I like the environmental aspect of the work, the people are friendly and I also like NERC as an organisation.

As to what step I take next, this is the grey area that I am not sure about yet. I have lots of options including a PhD, move into environmental education, and environmental protection officer. At the moment I do not wish to do a PhD as I need a break from study after the intensity of work during the Master's degree. I would like to take a side step from active research into pollution control with an organisation such as the Environment Agency at some point in the future, if and when the opportunity arises. I have also seriously considered applying for Voluntary Services Overseas (VSO) as they advertise for environmental scientists in various countries. Eventually I would like to move into environmental consultancy of some kind once I have adequate experience. The sky's the limit as they say!'

Choosing the right place

When choosing a degree course you have many things to consider: the syllabus, the university or college facilities, and the social and educational environments. There may also be important differences in approach between courses. For instance, if you are interested in the applications of the subject, go for a course with lots of project work.

There is another important factor that could influence your choice: the ratio of women to men on the course. Girls applying for a course in engineering or physics, for example, can expect to be in a small minority. But it is very important to choose your course and university carefully: there is a big difference in terms of ratios (and attitudes) between courses. Find out what the present gender ratios are and try to talk to women students who are already there. Ask if there are any women lecturers.

For boys too there will be gender considerations. A few might prefer to be on an all-male course, while many would feel happier with a more even balance of the sexes.

After graduation

Even in subjects where women 'hold their own' at undergraduate level, or are even in the majority, they begin to be outnumbered after graduating. For those pursuing a university career in the UK, the gender imbalance widens until, at very senior academic level, men predominate in most departments.

The Commission of Vice Chancellors and Principals has launched the Athena Project to focus on the unacceptable wastage of the skills and talents of women in Science, Engineering and Technology (SET) in higher education, including those who do not enter or consider entering careers in HE, and those who do, but get lost en route to the top. The project will offer incentives to HE institutions to improve the participation and progression of women in SET.

The gap widens

So, as people go through the education system from GCSE, to A level, to university and then on to postgraduate study, females are gradually outnumbered by males. Why does this happen?

There are all sorts of theories put forward to explain this phenomenon. Is it, for instance, something to do with a difference in the 'innate abilities' of males and females? Physics requires more maths than biology does and needs higher 'spatial' ability: that is the ability to deal with three dimensions and relationships in space. While this is true, the differences between males and females found in whole population studies, in spatial ability, are very slight. So it does not explain the huge discrepancy in the numbers of males and females choosing science subjects such as physics.

In other words, there are many girls choosing not to study maths and science even though their mathematical and spatial abilities are considerably higher than many boys who do choose the subjects. Modern biology requires a high level of spatial ability in order to understand molecular structures, and women deal with it just as well as men.

The Research Councils and the Wellcome Trust are co-funding research to explore whether there are differences in attitude between men and women in research careers. A 1997 Swedish study of success rates for male and female applicants for grants from the Swedish Medical Research Council identified a significant gender bias against female applicants. The UK Research Councils have examined their own success rates and found no evidence of gender bias, but there was anecdotal evidence that women were less likely to apply than their equally-qualified male colleagues.

Career perceptions

Another important factor influencing the choices that girls make is their perception of the careers that science subjects lead on to. The point was made earlier that there is an expectation that someone studying science will become 'a scientist'.

The perception of the biological sciences is that they lead to careers in healthcare or the environment: jobs which are about helping people and looking after the natural world. Careers in these areas seem to girls, and to many boys, to be more useful and socially acceptable when compared with physics, with its perceived links to weapons of war and causes of pollution.

> 'Physics will only lose its macho image when it is seen to employ resources more usefully.'
> *A woman physics teacher in a letter to a newspaper*

It is unfortunate that many girls rule certain subjects out without really considering the potential career development. Their decisions are often based on a limited view of the context in which these subjects are useful. Physics, for example, has many applications in communications, the media, health (e.g. optics, cancer diagnosis and treatment), the development of new materials, computers and the environment. Physicists are involved, with other scientists and mathematicians, in the bid to explain and tackle problems like the greenhouse effect and global warming.

Once again, appearances can be deceptive, and opinions often owe more to prejudice than fact. It will pay you to find out more before you make a decision.

Women at work

Women today are going into a much wider range of jobs. Changes in attitudes and lifestyles mean that more young graduate mothers are choosing to stay on in their careers. Career patterns are constantly shifting and developments, like the new company structures discussed in chapter 9, will benefit many women graduates.

Women science graduates are doing well in science-related career areas such as patent work (see profile page 139), information science, technical marketing, science journalism and publishing, and health and safety at work. Sadly, many girls are unaware of all these opportunities when they become discouraged and drop science early on in school.

Other areas where women with a science education and good communication skills are making an invaluable contribution are education and public understanding of science issues. Although they are in a tiny minority, women are slowly moving into politics and the higher echelons of the administrative Civil Service, where there is a crying need for a greater understanding of science and science issues.

In industry, scientists are being employed increasingly in teams of specialists. This is the type of co-operative working practice which many women (and men) prefer to the more hierarchical structure of the traditional company.

Finally, the transferable skills from a science degree, such as numeracy, logical and systematic problem-solving, and work planning, can be very powerful tools when combined with expertise in areas like communications that are often regarded as 'feminine skills'. The administration of science, for instance, is an increasingly important field, as funding has to be specifically targeted and resources carefully managed. People employed in research science have to explain precisely what they are trying to do and why, in order to convince grant-awarding bodies of the importance of their work; so women who write and communicate well will have the edge.

Discrimination

It is one thing to choose a science career, but quite another to succeed and advance within it. Unfortunately, discrimination occurs in all types of work, but it can be particularly difficult for a woman going into a company or a research group dominated by men who are suspicious of women graduates. In the USA recently, the Massachusetts Institute of Technology publicly acknowledged that it has been discriminating against women for years! Clearly some women can deal with it much better than others, but it is a very important factor to be aware of when considering a job, for exactly the same reasons as when choosing a degree course.

Women's careers are traditionally more likely to be affected by having a family than men's are. This may be why more female graduates choose teaching as a career either straight after university or later on. It might also explain why more women settle for posts as science technicians rather than going further. With the rapid pace of technological developments there is obvious concern that any career break could mean a woman in a science-oriented career feeling out of touch very quickly.

These pressures occur in other areas of work too, but research scientists in particular have to establish themselves when they are young. They must be prepared to work long and anti-social hours in the lab at a time when they may have very young families, or when they want to start a family.

It is not easy to do this or any other job within a two-career family partnership, but these are becoming more common and accepted. One big advantage of both parents establishing their careers is that two salaries certainly help with childcare costs.

How will the situation change?

The main thing that will help is more women studying science or becoming scientists. You will find that this is already happening - but slowly. There are more role models at school, at university and in work. There is also more pressure to give active encouragement to all young people who want to do science, but particularly to girls.

There have also been some improvements in schools. In response to concerns about young people no longer being interested in science, there is now a lot more imaginative teaching material for all ages. Exam questions and textbooks are now much more carefully scrutinised for gender bias and teachers (most of them anyway!) are more careful nowadays.

Your friends and classmates have a strong influence on you, and the pressure on girls to give up science doesn't all come from the opposite sex! But this is changing as more and more girls do well in GCSE science. There are lots of examples of girls helping each other, too. When women undergraduates find themselves outnumbered on university courses they often work together in practical classes and in tutorials to provide mutual support.

Scientist at work by Vicki

In the past few years the huge growth of interest in biological sciences, where women are better represented than in the other science subjects, has seen more women in higher academic jobs and in jobs in the growing biotechnology industry. According to the Biotechnology and Biological Sciences Research Council, women make up 45% of scientific officer applicants, mirroring the balance of women in the biological sciences; 60% of all those appointed are women.

As Europe opens up, the influence from other countries where women have a stronger position is increasing. More British science students than ever are studying or working on the continent as part of their education. This may be why British women postgraduates are now choosing to do their postdoctoral training in Europe rather than in the USA.

Support for women

As well as the informal help mentioned above and the strong support many girls get from their families and teachers, there are many women's groups dedicated to helping women choose and pursue a career in science. Women in Science and Engineering (WISE) is a national group with local branches which provides courses and speakers. Most of the professional organisations also have women's groups: Women Physicists, Women Engineers, Women Environmentalists, for example. All these groups offer advice and support either to individuals or to schools.

Every year universities and science organisations run short courses, usually lasting one week, for A-level students. They have titles like *Insight into Engineering*, *Chemical Engineering: a Career for Women*, *Girls into Science and Engineering*, *Science Workshop for Girls* and *Women and Physics*.

To encourage more women to take up science and technology, Coventry University is offering a free part-time course that could lead on to a degree. The expenses-paid access programme offered by the university's school of natural and environmental sciences offers free childcare and travel, and waives fees. No qualifications are needed to enrol. The 18-month course is backed by a European grant.

Many companies are supporting Opportunity 2000, the campaign to improve women's employment opportunities. In 1996, women had already secured one third of middle management and 41% of junior management positions in those member organisations. Glaxo Wellcome is one such company. Of 3500 staff, nearly two-thirds hold scientific roles, and almost 50% of them are women. Women's representation amongst senior staff managers and research managers has increased from 10% to 18% over the last four years.

Science needs more women

Don't be discouraged from studying science: the idea that it is a man's subject has had its day.

- ☐ Science needs more women as much as women need the career opportunities that a science degree can offer.
- ☐ Girls are now doing better than boys at GCSE in sciences. They can do the same at A level.

- If you are not sure about science, choose your A levels so that you keep your options open as long as you can: it's very hard to go back.
- A science degree does not just lead into a laboratory: there are lots of other careers open to science graduates.
- There are lots of interesting women science graduates like the ones in the profiles in this book. Try to meet and talk to some.
- There are many organisations that can help and give you support. Make good use of them.

In summary

Many girls are opting out of science in spirit long before they reach GCSE. They are influenced at a time when peer group pressure is strong and are easily discouraged if they find the subjects difficult. They decide against science before they are aware of the wide range of careers that science qualifications can offer.

Women seem to prefer biology to physics and tend to choose subjects where they can see applications in contexts such as health or the environment. Women who do take maths and physics are more likely than their male peers to use their qualifications in teaching.

Fewer women science graduates than men opt for postgraduate study, apart from teacher training.

The practicalities of family life still seem to have a strong influence, but this will change as more young women graduates combine a career with having children, and employers introduce more flexible working practices for their graduate staff.

There are many welcome changes ahead; as career structures are altered, as more women gradually become established in science-based careers, and as more enlightened practices from the continent begin to influence men and women in Britain. Things would change faster if there were more young women science graduates around to help!

Spot the female scientist! (Answers)

Dorothy Hodgkin (1910-94) - crystallographer who discovered the structure of vitamin B12.

Careers with a Science Degree

Marie Curie (1867-1934) - physicist who discovered radioactivity and won two Nobel prizes.

Barbara McClintock (1902-92) - geneticist who won the 1983 Nobel prize for Physiology and Medicine.

Rosalind Franklin (1920-58) - chemist and crystallographer who made a major contribution to the discovery of DNA.

Caroline Herschel (1750-1848) - astronomer and mathematician.

Chapter 4: Science courses in higher education

This chapter covers:
- ☐ the variety of science courses available
- ☐ the links between science qualifications at school, college and university
- ☐ the importance of mathematics
- ☐ how you can get into science with non-science A levels (or equivalent)
- ☐ how you can keep open a range of options
- ☐ where to go for further information and advice.

How do you decide?

One big advantage of taking science A levels, Highers, BTEC or GNVQ qualifications is that you will have a huge range of courses in higher education to choose from. The downside is that this choice can be quite bewildering.

The first questions to ask yourself

- ☐ Do I want to build on the subjects I am doing at present and look at courses where these sciences are requirements for entry?
- ☐ Shall I look at courses where some scientific knowledge and interest is required, but which take people with a wide range of backgrounds?
- ☐ Should I consider courses which will accept students with any A level subjects or equivalent, even those who have not studied sciences since GCSE?

The important thing here is to remember the one-way escalator idea introduced in chapter 1: it is much easier to become less scientific than more scientific. It is much easier to become less mathematical than more mathematical. If you have A levels in non-science subjects and would like to switch to science, it is not too late (see 'What about changing track?' later in this chapter). There are science degree and diploma courses that will accept applicants with any A level subjects. However, if you choose to go for a degree in a subject such as law or business studies it is very unlikely that you will be able to get back into sciences later on.

Maintain your mathematics base

Because mathematics is such an essential tool for sciences it is important to try to take it as far as you can, even if you are not considering maths at A level or Highers. So persevere at GCSE and try to take the higher tier papers rather than intermediate. That way you won't miss out on important concepts which will be useful if you decide to do science A levels. If you are not sure whether you are capable of taking a full maths A level, remember that the first year of the course will stand as an AS qualification if you decide not to continue. Discuss the options with your teachers. Remember that the further you progress in mathematics, the more it becomes a language of ideas and less a matter of doing calculations.

Mathematics at A level offers several different syllabuses. You should discuss with your teachers what is available in your school and which option would be most useful for your A level combinations and future career.

If you are taking a science-based BTEC National or Advanced GNVQ or a Scottish National Qualification, you should try to enhance it with as much extra mathematics as you can cope with. On a GNVQ course you have time available for additional studies, so mathematics at some level could fit in here.

Courses that require sciences at A level or equivalent

First, there are courses in subjects you are very familiar with from school - the basic sciences:

- ☐ mathematics
- ☐ physics
- ☐ chemistry
- ☐ biology.

Then there are science subjects you might have come across at school or college. These do not require an A level in that particular subject for acceptance onto a degree course, although A levels or equivalent in other sciences may be required. These include:

- geology (or earth sciences)
- environmental science
- computer science
- psychology.

In addition, there are many courses offering combinations of these eight subjects and these subjects together with a variety of different ones.

Different types of course

You may want to consider different levels of study, degrees or Higher National Diplomas, and compare full-time courses with sandwich courses. Would you rather do a single subject or combined course? Have you thought of spending time getting work experience as part of your studies? What about studying in another country as part of your course? These are a few of the many questions you need to ask yourself. It's your future. Only you can make the right choice.

In England and Wales, most honours degree courses are usually three years long, although there are also four-year enhanced science and engineering courses. Sandwich degrees last for four years. In Scottish universities, honours degree courses last for four years.

Full-time Higher National Diploma courses usually take two years, or three years as a 'sandwich', where periods of full-time study alternate with periods of practical experience in a working situation. Diploma courses are different in approach from degree courses, and include more project-based work and less discrete subject teaching. It is sometimes possible to move from an HND course to a degree course in a similar subject after one or two years of study, especially if that course is at the same college or university.

The minimum qualifications for degree courses are normally two A level passes, an Advanced GNVQ (often with additional modules or an A level) or a BTEC National Diploma, plus five GCSEs at grades A-C. Adults with relevant knowledge and experience may be accepted with fewer qualifications, or after an Access course. For Higher National Diploma courses the entry requirements are flexible, but likely to be of the standard of the equivalent of at least one A level or NVQ level 3.

Degree subjects

Subject titles do not always give a clear idea of what the courses are about. Make a list of the ones that interest you and find out more about them. The glossary on page 167 describes some of the scientific subjects offered as degree and diploma courses. It will help you to choose. You should also refer to prospectuses, the UCAS Official Guide to University and College Entrance and the CRAC Degree Course Guides for each subject. You'll find other useful books in the book list on page 163. On the next few pages there are lists of subjects to give you an overview of the range of disciplines available to you.

They have been categorised as follows:

- specialist science subjects, most of which you may have touched on at school
- the main engineering subjects
- the specialist engineering subjects
- specialist technology subjects
- medically-related courses
- courses relating to agriculture
- courses for which some science knowledge is useful.

Please note that these lists are by no means fully comprehensive, although they do cover most subject areas. New subject combinations and new courses with new titles are appearing every year. You must get the most up-to-date information before you make your choice.

N.B. Many of these subjects are available as Higher National Diploma courses as well as degree courses.

Specialist science subjects, most of which you may have touched on at school

- Acoustics
- Anatomy
- Artificial intelligence
- Astronomy
- Astrophysics
- Biochemistry

Chapter 4 - Science courses in higher education

- ☐ Biophysics
- ☐ Biotechnology
- ☐ Botany
- ☐ Computer science
- ☐ Cybernetics
- ☐ Earth sciences
- ☐ Ecology
- ☐ Electronics
- ☐ Environmental science
- ☐ Ergonomics
- ☐ Food science
- ☐ Genetics
- ☐ Geochemistry
- ☐ Geography (if a BSc)
- ☐ Geology
- ☐ Geophysics
- ☐ Immunology
- ☐ Marine biology
- ☐ Materials science
- ☐ Metallurgy
- ☐ Microbiology
- ☐ Molecular biology
- ☐ Nutrition
- ☐ Oceanography
- ☐ Parasitology
- ☐ Pathology
- ☐ Pharmacology
- ☐ Physiology
- ☐ Polymer science
- ☐ Space science
- ☐ Statistics
- ☐ Virology
- ☐ Zoology

Again these subjects can often be taken in combination with a basic science subject or with each other. They could even be studied together with a completely different subject such as marketing, computing, music, a foreign language or business studies.

Then there are all the engineering and technology subjects which will normally require maths and physics at A level or equivalent.

The main engineering subjects
- ☐ Chemical engineering
- ☐ Civil engineering
- ☐ Electrical engineering
- ☐ Electronic engineering
- ☐ Mechanical engineering
- ☐ Production engineering

Specialist engineering subjects
- ☐ Acoustics engineering
- ☐ Aeronautical engineering
- ☐ Aerospace engineering
- ☐ Agricultural engineering
- ☐ Air transport engineering
- ☐ Biochemical engineering
- ☐ Building services engineering
- ☐ Communications engineering
- ☐ Computer engineering
- ☐ Control systems engineering
- ☐ Design engineering
- ☐ Digital systems engineering
- ☐ Environmental engineering
- ☐ Fuel and energy engineering
- ☐ Industrial engineering
- ☐ Marine engineering
- ☐ Materials engineering
- ☐ Mining engineering
- ☐ Motor vehicle engineering

- Multimedia engineering
- Naval architecture
- Petroleum engineering
- Plant engineering
- Power engineering
- Structural engineering
- Systems engineering
- Telecommunications
- Video engineering
- Yacht and powercraft design

For this group of specialist courses you really do have to find out if the course is what you think it will be and make sure you investigate the employment prospects thoroughly. This is particularly important because specialist courses like these are usually aimed at employment in a very specific industry, and are less likely to be acceptable in a wide range of other jobs. Don't forget you can enter engineering with degrees in other relevant disciplines. Paul Pilkington (page 89), now working in aeronautical engineering, says that his physics degree course 'covered all the major topics of the engineering degrees, but in less detail, but also covered subjects like astrophysics and cosmology and particle physics'.

Specialist technology subjects

- Brewing and distilling
- Building technology
- Ceramics science
- Food technology
- Glass science
- Internet technology
- Laser technology
- Leather technology
- Medical electronics
- Television technology
- Textile technology
- Timber technology
- Water science

Again these choices are very specialised and would not be a good choice for someone keen to keep a wide range of options open! Take care to check the employment opportunities carefully.

If you are taking BTEC National, Advanced GNVQ or Scottish National Qualifications in vocational subjects, it is important to check that the mathematical content of your course, including additional studies, will be sufficient for entry to the higher education course you have in mind.

Medically-related courses also leading to professional qualifications

- Chiropractic
- Dentistry
- Dietetics
- Environmental health
- Medicine
- Nursing
- Occupational therapy
- Optometry
- Orthoptics
- Osteopathy
- Paramedic science
- Pharmacy
- Physiotherapy
- Podiatry
- Prosthetics and orthotics
- Radiography
- Radiotherapy
- Speech and language therapy
- Sports therapy
- Veterinary science

A few of these courses may also accept applicants with non-science A levels or equivalent.

There are HND courses in subjects related to medicine, but these are not professional qualifications.

Courses in both medicine and veterinary science will include many of the subjects listed on page 68, such as anatomy, physiology, biochemistry, pharmacology and pathology, in addition to clinical studies.

Courses relating to agriculture

- ☐ Agricultural economics
- ☐ Agriculture
- ☐ Animal production
- ☐ Arboriculture
- ☐ Crop science
- ☐ Estate management
- ☐ Fishery science
- ☐ Forestry
- ☐ Horticulture
- ☐ Land surveying
- ☐ Poultry production and management
- ☐ Rural resource management
- ☐ Soil science
- ☐ World agriculture

Higher National Diploma courses are also available in most subjects related to agriculture.

Courses for which some science knowledge is useful, although they are open to students with other entry qualifications

- ☐ Anthropology (physical)
- ☐ Archaeology
- ☐ Architecture
- ☐ Beauty therapy/management
- ☐ Behavioural sciences
- ☐ Built environment studies
- ☐ Cartography/mapping science
- ☐ Catering management
- ☐ Cognitive science

Careers with a Science Degree

- ☐ Computer studies
- ☐ Geographical information systems
- ☐ Health studies
- ☐ History of science
- ☐ Home economics
- ☐ Information management
- ☐ Landscape architecture
- ☐ Leisure and recreation studies
- ☐ Linguistics
- ☐ Photography/lens and digital media
- ☐ Printing management
- ☐ Psychology
- ☐ Sports studies
- ☐ Surveying
- ☐ Town and country planning

Your range of options from a new angle

If these lists seem difficult to use, why not try another way of considering the subjects that might interest you in higher education? Start with your favourite basic science subject at A level and see what the option list looks like.

Many of the maths-based courses will also require physics - and vice versa; many of the courses requiring chemistry will also require another science, and those requiring biology may well also need chemistry. CHECK CAREFULLY FOR EXACT ENTRY REQUIREMENTS.

If your favourite subject is MATHEMATICS the following courses might appeal to you:

- ☐ Actuarial science
- ☐ Aeronautical engineering
- ☐ Aerospace studies
- ☐ Agricultural engineering
- ☐ Artificial intelligence
- ☐ Chemical engineering

☐ Civil engineering
☐ Computer science
☐ Control systems engineering
☐ Cybernetics
☐ Electrical engineering
☐ Electronics
☐ Management science
☐ Materials science
☐ Mathematical engineering
☐ Mathematics
☐ Mechanical engineering
☐ Meteorology
☐ Naval architecture
☐ Operational research
☐ Physics
☐ Quantity surveying
☐ Software engineering
☐ Space science
☐ Statistics
☐ Transport engineering.

If your favourite subject is PHYSICS these subjects might suit you:

☐ Acoustical engineering
☐ Aeronautical engineering
☐ Aerospace studies
☐ Agricultural engineering
☐ Astrophysics
☐ Biomedical engineering
☐ Biophysics
☐ Brewing
☐ Building technology
☐ Chemical engineering
☐ Civil engineering

Careers with a Science Degree

- Communications engineering
- Computer science
- Control systems engineering
- Cybernetics
- Dentistry
- Design engineering
- Earth sciences
- Electrical engineering
- Electronics
- Environmental engineering/ science
- Ergonomics
- Fuel and energy engineering
- Geology
- Geophysics
- Materials science
- Mathematical engineering
- Mathematics
- Mechanical engineering
- Medical physics
- Meteorology
- Naval architecture
- Oceanography
- Optics/optical sciences
- Optometry
- Orthoptics
- Photography
- Physics
- Physiotherapy
- Polymer science
- Quantity surveying
- Software engineering
- Space science
- Statistics
- Transport engineering.

If your favourite subject is CHEMISTRY you might like to consider these subjects:

- ☐ Biochemistry
- ☐ Biology
- ☐ Biomedical science
- ☐ Biotechnology
- ☐ Brewing
- ☐ Chemical engineering
- ☐ Chemistry
- ☐ Colour chemistry
- ☐ Dietetics
- ☐ Digital imaging sciences
- ☐ Ecology
- ☐ Environmental health
- ☐ Food science
- ☐ Geochemistry
- ☐ Immunology
- ☐ Materials science
- ☐ Medical biosciences
- ☐ Medicine
- ☐ Metallurgy
- ☐ Microbiology
- ☐ Mineralogy
- ☐ Molecular biology
- ☐ Nutrition
- ☐ Paper science
- ☐ Pathology
- ☐ Pharmacology
- ☐ Pharmacy
- ☐ Physiology
- ☐ Polymer science.

Careers with a Science Degree

If your favourite subject is BIOLOGY the following subjects might appeal:

- ☐ Agriculture
- ☐ Anatomy
- ☐ Anthropology
- ☐ Arboriculture
- ☐ Bacteriology
- ☐ Biochemistry
- ☐ Biology
- ☐ Biomedical science
- ☐ Biotechnology
- ☐ Botany
- ☐ Brewing
- ☐ Chiropractic
- ☐ Dentistry
- ☐ Dietetics
- ☐ Ecology
- ☐ Environmental health
- ☐ Food science
- ☐ Forestry
- ☐ Genetics
- ☐ Health sciences
- ☐ Horticulture
- ☐ Immunology
- ☐ Landscape architecture
- ☐ Marine biology
- ☐ Medical biosciences
- ☐ Medicine
- ☐ Microbiology
- ☐ Nutrition
- ☐ Occupational therapy
- ☐ Oceanography
- ☐ Optometry

- Orthoptics
- Parasitology
- Pathology
- Pharmacology
- Physiology
- Physiotherapy
- Plant science
- Podiatry
- Speech therapy
- Sports studies
- Veterinary science
- Zoology.

Choosing which A levels to do

Yet another approach might be helpful. Consider the four basic science subjects:

- mathematics
- physics
- chemistry
- biology.

To keep open the widest choice of science and other courses in higher education it would be best to take all four after GCSE! Four full A levels, however, are definitely too much for most people. Taking some subjects at A level and some at AS is an option, but this will not give the subjects equal weight. Deciding which to carry on to A level at the end of year 12 needs a lot of careful thought; you need to find out whether you can start three sciences and maths at AS and then continue with two or three to A level at your school or college. The new post-16 qualifications are supposed to encourage a broader range of subjects, so this might not be seen as a good use of the system.

If you are in Scotland, Highers will give you a broad base. You can also take BTEC National or Advanced GNVQ science for a more voactional slant and still be able to apply for a degree course - especially if you take additional studies in mathematics or a single science A level.

Here are some examples to consider

In theory, you can do up to five subjects in year 12 under the new post-16 system of qualifications; in practice some schools may not be able to offer more than four subjects because of timetabling restrictions. It may also be possible to pick up a new AS level or single award Advanced GNVQ in year 13. Some possible combinations are:

Year 12

Two sciences, maths and two non-science AS levels (or single award Advanced GNVQ in place of one subject)

or

Advanced GNVQ science plus maths and one or two non-science AS levels

Year 13

End study of one or two subjects, gaining AS level (or single award Advanced GNVQ)

and

Continue studying two sciences and maths or three sciences at A level

or

Continue studying two sciences plus a non-science subject at A level

or

Continue studying Advanced GNVQ plus maths A level

- ☐ Your actual choices will depend on what is on offer at your local schools and colleges of further education.
- ☐ A three-unit part GNVQ may be introduced in the future as the equivalent of an AS level.

Paul Compton (page 109) did not decide on scientific A levels until his GCSE results prompted him to move in that direction, having kept his options open to veer towards design or architecture instead.

You could take a BTEC National course in sciences

Although this leaves open a wide range of sciences you might need to do extra mathematics to reach A level equivalence in that subject. A BTEC National leads naturally on to a Higher National Diploma, and is also acceptable for many degree courses - although some still prefer traditional science A levels.

Narrowing down your choices

If keeping all your science options open seems too difficult, it may be more sensible to examine your interests in the basic sciences the other way round and look at the consequences of not taking each one.

Here are some guidelines

No mathematics?

This would eliminate most courses and careers in maths, physics, engineering and technology. Maths is also useful to support the other sciences at A level and in higher education, and would be helpful for other degree courses such as economics and management sciences.

If you are quite sure you don't want to go into the more mathematical areas and feel that your interests lie more on the biological and biochemical side, then continuing with maths might not be so important. On the other hand, you might need mathematics later on and you would be wise to study it for as long as you can. AS mathematics or statistics could be worth considering. It is easier to pick up both statistics and biology on your own than it is to make up lost ground in maths if you find you need it later on.

No physics?

This would limit your choice of courses and careers in physics, engineering and technology and in mathematical subjects. There are other careers such as optometry, acoustics and meteorology where physics is involved. If you want to do biological or chemical subjects then physics may not be so badly missed, particularly if you have an appropriate level of mathematics. However, physics does support chemistry and would help you with the molecular aspects of biology.

No chemistry?

If this leaves you with both maths and physics, you still have the whole range of mathematical, physics, engineering and technology careers to choose from. Some students take double maths and physics, or maths and physics with another arts or social science subject, such as economics. However, without chemistry, you would be disadvantaged in fields such as chemical engineering and other biological and biochemical subjects for which chemistry is more important than biology. Dropping chemistry is a mistake if you are interested in medicine, veterinary science or pharmacy. You would find yourself in trouble.

No biology?

Whilst biology is the most popular scientific A level after maths, it is usually regarded as the least 'hard' of the sciences. Quite what this means is rather unclear. Possibly it is because biology is a more descriptive science and many people find the concepts easier to grasp than those in physics. But this is changing as the study of biology becomes less about whole organisms and more about what goes on at the molecular and cellular levels. As mentioned above, people find it easier to study biology on their own and catch up with the subject later on, say at degree level, than if they had missed out on other basic sciences. Biology is very rarely a requirement for entry to degree courses (even for biology!), although it counts fully as a science subject if the entry requirement is for two unspecified science subjects.

People who are interested in a biology-related career often include biology in their A level choice because it seems more relevant than the other basic sciences to their ultimate goal, say medicine or agriculture, although biology is actually not essential for either of these. So if you decide to study biology in place of another science, you have to be clear about what future courses and careers you are eliminating by the choice you make. If you think you might want to study biological subjects beyond A level, you don't have to take biology at A level, but make sure chemistry is included because it will be essential for the next stage.

Remember that there are opportunities to catch up by taking a foundation or extra year on those degree courses which accept applicants without science A levels.

What about two basic sciences at A level?

The sections above have looked at ways of including as many sciences as possible as well as maths in your A level choice in order to keep open the widest range of options. But many students take only two or even one of these subjects in their A level and AS combinations. This is often because they want a wider range of subjects at A level, such as two sciences with a humanities subject, or two sciences with a social science subject. You might also decide to take a different science subject such as environmental science or computer science because it interests you and you feel it would complement your other subjects.

Taking just two basic sciences will clearly narrow your choice at the next stage, but this will not matter if the careers and further courses being eliminated are not ones that you are likely to be interested in. For example, a combination of maths, physics and economics leaves open a wide range of maths, physics and engineering options, but cuts out courses requiring A level chemistry (such as medicine and biochemistry). An A level choice of chemistry, biology and French would make it difficult to get into engineering and physics-related careers and also those requiring a high level of mathematics.

Again, the best way of reviewing the career options leading from any particular choice of two basic sciences is to look at the careers and further courses being eliminated, or made less easy to enter, by the combination you are considering. You can use the lists above to help you.

Is it worth taking one basic science?

It is not usually a good idea to take either physics or chemistry on its own, unless you are just interested in widening your A level choice for general educational reasons rather than studying sciences beyond A level. The most usual single subjects are mathematics (which is useful for economics and business courses) and biology (which is a good foundation for paramedical studies or for those who might find it useful later on, in primary teaching, for example).

A single science A level is really too narrow a base for most science degrees. However, now that many science courses in higher education (apart from the ever-popular medicine and veterinary science) are keen to get more applicants, they are becoming more flexible about entry requirements. Supporting AS levels will count towards your entry requirements, and maths and science GCSEs at grades A-C may also widen your choices.

What grades will you need?

This section has been deliberately left to the end because choosing subjects on the basis of grades required rather than following your scientific and career interests is not likely to end in success. However, it is also sensible to be realistic: there are some courses, such as medicine and veterinary sciences, where the competition is such that high grades are needed to get in.

Very often the grades required for a degree course depend more on the popularity of the course rather than on the difficulty of the subject. So there are huge variations between universities and colleges in terms of the grades they require for the same subject, as well as for different subjects. There are exceptions: although mathematics and physics, for example, have never had a very high ratio of applicants to places - in fact some courses have been discontinued because of lack of take-up - these courses do attract students with high grades. This is possibly because students feel these subjects would be difficult, or because they have been advised against doing them. In contrast, business studies is an example of a subject which attracts high numbers of applicants who, on average, do not have very high grades. One or two of the people in our profiles did not get good enough grades for their first choice of course, but ended up on courses which they thoroughly enjoyed and have been the basis for successful careers.

Scientist at work by Rebecca

Information about the grades likely to be attached to offers of places on degree courses is published in the *UCAS Guide to University and College Entrance*, and also in *Degree Course Offers*, published annually by Trotman. Remember, this information may help you to look at your applications realistically but should not be used as the sole basis for your choice.

As higher education absorbs more students, there is now a wider spread of offers. There is a trend towards valuing communication and teamwork skills and other non-academic attributes, as well as exam passes, when offering places. Many colleges of higher education will accept students with the minimum two A level passes onto degree courses. However, the table below shows that point scores have increased since 1994, reversing a downward trend in some subjects of the previous few years.

Competition in the long-established universities is generally the keenest, and grades normally play a very significant part in selection, although at present there is a shortage of students with A levels in sciences.

For applicants accepted by institutions in the UCAS scheme over the past few years, the league table of A level points is on the page opposite. When reading this table remember that:

- within these groupings there are subjects that are a lot less popular than others. Law, for example, requires grades almost as high as medicine, whereas grades for other social sciences tend to be much lower
- these are average grades and there are many degree and diploma courses, both at older and new universities, that will take students with lower grades. Others, on the other hand, including Oxford and Cambridge, will require very much higher grades for all subjects
- the points tariff is changing for entry to degree courses in autumn 2002.

Average points required by institutions in the UCAS scheme

	1994	1996	1998
Medicine and dentistry	26.3	27.9	28.5
Social studies	19.0	19.4	19.8
Languages and related disciplines	20.5	21.3	21.7
Humanities	19.7	20.4	21.3
Business and administrative studies	15.7	16.8	17.3
Mathematical sciences and informatics	17.1	18.9	18.8
Architecture, building and planning	15.3	16.5	17.0
Creative arts	16.9	17.5	17.2
Physical sciences	17.3	19.0	19.8
Biological sciences	18.2	19.7	20.3
Agriculture and related subjects	17.5	19.1	19.5
Subjects allied to medicine	17.5	18.9	18.7
Engineering and technology	16.4	18.9	19.5
Education	13.5	14.5	15.2
Combined sciences	16.8	19.1	19.8
ALL SUBJECTS	17.6	18.8	19.2

Source: UCAS

Points tariff

A level scores
A = 10 pts
B = 8 pts
C = 6 pts
D = 4 pts
E = 2 pts

AS level scores
A = 5 pts
B = 4 pts
C = 3 pts
D = 2 pts
E = 1 pt

Grades required by BTEC holders

For entry to degree courses, students with BTEC National qualifications will usually be expected to achieve several Merits and, for some courses, Distinctions too. Again, details are available in the *UCAS Guide to University and College Entrance* and *Degree Course Offers*.

GNVQ and GSVQ applicants

After a long period of familiarisation, admissions tutors are mostly now well acquainted with these qualifications. (In Scotland, the GSVQ is being phased out and replaced by national units or courses.) The *UCAS Guide* and *Degree Course Offers* also include information about their acceptance. Many science-based degree courses will require a science or maths A level in addition to the GNVQ or GSVQ, or at least additional units. Advanced GNVQs in Engineering, Information Technology or Manufacturing may be acceptable for some vocational degree courses, besides Science.

Combining sciences with other subjects

Many people choose a mixture of A levels by studying social science, humanities or creative subjects alongside science. This can work well; it can give you a more varied programme and will allow you insights into a range of very different subjects. However, it will cut down the range of higher education and career options ahead, the same as when you reduce four sciences to three, only more so. Think carefully, in case you decide to continue with science later on. Taking other subjects at AS level only is one way of broadening your subject base without minimising your options.

Celia Keen (see page 139) studied German alongside chemistry, physics and maths 'for sheer pleasure', but it turned out to be extremely useful in her career as a Chartered European Patent Attourncy.

Jessica Harris (see page 120) studied sociology and economics with mathematics, which are both relevant to her present post as a medical statistician.

In Scotland, where students have been able to choose five subjects for Highers, most students wanting to do sciences have been encouraged to do at least one social science or humanities subject.

What about changing track?

If you are quite sure you want to leave science behind there are many courses which will take students with any A level or equivalent qualification: accountancy, economics, philosophy, social science, business studies, law and librarianship to name just a few. For some of these courses mathematics at A level or equivalent will be required.

The range of non-science degrees open to people with either science or any A levels is considerable, but certain science courses, like computer sciences, sports science and environmental studies, may accept people with A levels in other subjects. There are also science, engineering and technology courses where students without science A levels or equivalent can complete a foundation course or a preliminary intensive science year before going on to a three-year degree course alongside those with science qualifications. Mathematics at GCSE grade A-C is usually required.

A few medical schools also offer one-year introductory courses for students with non-science A levels or equivalent qualifications. These courses are extremely competitive and usually require very high grades and evidence of strong motivation.

Full details of all these courses will be in the *UCAS Handbook* and other publications in the Book List (page 163).

What help can you get?

- ☐ Use the careers and higher education information and guidance resources listed in Chapter 10. They may be available in your school, college or careers centre library.
- ☐ Talk to your science teachers.

- ☐ Read more about the courses on offer in prospectuses, higher education handbooks and databases.
- ☐ Widen your general science knowledge through magazines such as *New Scientist*, the science pages of the quality newspapers, 'popular' science books and biographies (see page 164), TV and radio programmes, visits, lectures, open days and 'taster' courses at universities.
- ☐ Visit the websites of the major scientific institutions.
- ☐ Talk to your careers adviser.
- ☐ Talk to someone who has completed a course that you are interested in, if you can.
- ☐ Ask local employers.
- ☐ Try some work experience or holiday jobs.
- ☐ Discuss your thoughts with family and friends.

Garry Hepherd

Garry's experience shows how your career can flourish, even if your A levels and degree choice are not quite what you originally intended!

Career profile

Age: 36

Account Manager, Rigid Foam, ICI Polyurethanes

A levels: chemistry, general studies

Degree: BSc (Hons) 2:1 Textile Chemistry, University of Leeds

'I never really had a fixed career path in mind whilst at school (which has proven to be just as well really given today's demanding and continually changing workplace) and my career has evolved and developed in response to market and company needs and personal ambitions.

I was educated at a traditional grammar school in the 1970s and was forced to make a choice between a science or arts biased subject route at an early age. I found this dogmatic approach to learning particularly limiting since I was not allowed to combine my three favourite subjects - chemistry, history and French - at A level and, being particularly strong in chemistry (I was also entered for S level chemistry), was recommended, and chose, to take chemistry, physics and maths as my A level subjects, together with general studies.

By the second year of A level study I had decided that I would like to study chemistry at university, with a vague idea that this may lead to a career in teaching, but the only problem would be that this would require good passes in all three sciences. Not surprisingly, I did not pass either maths or physics but having a strong interest in politics and current affairs I did manage to get a D grade in general studies and, of course, a B grade in chemistry.

It was at this stage in my life that I received some excellent career advice from my headteacher. I was faced with three options :

- Resit A levels and try to secure a university place to study chemistry

- Take up an offer to study HND Chemistry at Plymouth polytechnic

- Take up an offer to study BSc Textile Chemistry at Leeds University

I did not relish the thought of retaking maths and physics A levels! I knew that I wanted to leave home and experience university life, but, originating from Birmingham, I did not have any experience of textile manufacture nor the textile industry. My relatively poor performance at A level was not helping me to expand my options beyond the above three choices via clearing, and thus I was advised by my headteacher that I should take up the Leeds offer.

The Department of Textile Industries at Leeds University (like every other school of textiles in the UK) always had problems in filling their Textile Science and Technology courses - Textile Design on the other hand was always massively oversubscribed - and they often attracted people onto the course by selecting candidates, like myself, who had originally applied to the School of Chemistry at Leeds but had been unsuccessful in attaining a place.

My headteacher, knowing how the academic system worked, advised me to work hard in my first year, obtain good results in the university exams and then switch courses at the end of year one (the Textile Chemistry course had a high pure chemistry content and involved lectures/teaching within the school of chemistry). This, he reasoned, would be the fastest route to a degree in chemistry and far easier than resitting my A levels.

To my surprise and delight, I found the Textile Chemistry course interesting , stimulating and enjoyable. So much so, that I transformed within one year from a 'failed' A level student to the top student in the year end exams and so, like the previous summer, I was now faced with more choices.

As predicted by my headteacher, based on my first year exam performance, the School of Chemistry was now willing to accept me on the Pure Chemistry course, but I was also fortunate enough to receive a further piece of help and advice from my university tutor at this important time. Major textile manufacturers and retailers, such as Courtaulds plc and Marks & Spencer, were facing difficulties in recruiting well-qualified textile scientists and technologists to the industry and had launched a sponsorship scheme to attract graduates into the textile industry.

My university tutor introduced me to the Courtaulds recruitment manager, and after several interviews I secured a place on the Courtaulds graduate sponsorship scheme. The agreement meant that I received financial assistance for books etc from Courtaulds and, in return, I worked for them during university vacations with a view to joining them full-time upon graduation, subject to my attaining a satisfactory degree. An additional bonus was that I was also invited to join Courtaulds management training courses during my vacation periods.

Thus my first permanent job was secured at this early stage in my university studies and motivated me to settle in the textile department (which I was really enjoying). I also became active in the textile society, eventually being elected treasurer.

I left Leeds University in 1984 with 2:1 BSc(Hons) in Textile Chemistry and duly joined Samuel Courtaulds Fashion Fabrics (a division of Courtaulds Textiles) as a management trainee based at Wharf Mill in Hyde, Chesire. After a period completing projects in a number of departments, I was asked to take charge of a team of people working a three shift rotation pattern (6-2, 2-10, 10-6) in the weaving preparation department. This was a very challenging but yet enjoyable experience and, although difficult at times, helped me to develop my human-resource management, leadership and teamwork skills.

Whilst working in this supervisory position, I noted that Albright & Wilson - a Midlands chemical manufacturing company - were looking to recruit textile chemists for their flame retardants business. I applied for this position because it would allow me to make better use of my academic background on a daily basis. The business was international and provided opportunities for overseas travel - which greatly appealed to me - and the job was based in my home town - Birmingham.

I joined Albright & Wilson in 1985 as a Technical Service Chemist. The job involved both internal laboratory work as well as carrying out training courses and supervising technical trials at customer premises throughout Europe. This was valuable experience in developing my communication and presentation skills.

After about 18 months in this technical position an opportunity arose to move into a commercial role, and I was appointed Technical Sales Executive of flame retardants, which involved direct selling of a range of speciality chemicals to the textiles, polyurethanes, timber and paint industries throughout the UK and Eire. My technical background helped in selling *these speciality chemicals, and the job allowed me to develop my commercial and business skills and, after trying production and technical service jobs, I realised that I had finally found the job role which suited me.*

During my 12 years with Albright & Wilson I had to adapt to a number of internal reorganisations and a change of company ownership, and I had to be flexible enough to take on different export sales responsibilities and challenges, culminating in my appointment to European Sales Manager with overall commercial responsibility for a business unit and sales team.

After this most enjoyable and rewarding period of my career, an opportunity arose to join a division of a newly formed textile group - International Performance Textiles (IPT) as Business Manager for their knitwear yarns product range. This job appealed to me because IPT were looking to recruit a number of new young managers into their company with a view to trying to turn around a number of traditional textile mills. I joined the company in the summer of 1997. My job was to market and sell a new range of total easy care (machine washable plus tumble dry performance) worsted spun lambswool yarns, to Marks & Spencer and more specifically to their knitwear suppliers.

Although we successfully developed new business with these products, the last two years have been particularly difficult for the UK textile industry, and 18 months after joining the company, IPT announced that they would have to close a number of UK mills and in September of 1998 I was made redundant.

To their credit, IPT provided support to try and help people find alternative employment and I was able to get career counselling advice from a consultancy company who helped me to prepare my CV, identify and research potential new employers and improve my interview technique. It was, however, through contacting friends and former colleagues that I heard that ICI were looking to recruit an account manager for their polyurethanes business and I applied and successfully secured the job in December 1998.

Although my period of unemployment was frustrating, the structured approach to job searching that I learnt from the recruitment consultants certainly kept me busy and reinforced the value of networking and maintaining contacts in business life.

Thus some 15 years after leaving university, I have been fortunate enough to have had a most varied, challenging but enjoyable career. It probably has not been the most structured career path that you will find but, as I stated at the beginning, today's employment world is constantly changing and you must be flexible and acceptant of change if you are to survive.'

A final word

This chapter is just a starting point. There are many other things to be done before you can reach a decision. There are new ideas to consider, people to talk to and other sources of help and information to explore. The lists earlier in this chapter do not cover every option: there are new courses offered every year and all sorts of combinations you will never have heard of that may be available at just one or two institutions.

So you must consult the most up-to-date reference books about higher education, particularly *The Official UCAS Guide to University and College Entrance* and the others in the Book List on page 163. Study the Glossary on page 167. It is most important that you find out what the science courses in higher education are about, particularly if they are subjects you have not studied at A level. Don't just assume that you know: make sure you understand what is involved.

Consider the stage you are at now, and decide where you are in your career planning. Draw up a plan of action using the suggestions in this book and then add your own. For example, you might want to talk to someone who is already in a job that interests you. The more you find out, the more your interests will clarify and new ideas will grow. You want to make sure that you choose the right course, the one which will suit you best, because that is how you will gain success and satisfaction.

Chapter 5: Choosing and working with your science degree

This chapter includes:
- ☐ some of the criteria you might employ in choosing a science degree
- ☐ the three main ways that people use their science degrees in jobs
- ☐ examples of the transferable skills that science graduates can use in any job
- ☐ profiles of young scientists who have used their degrees in different ways.

Choosing a degree or diploma course

All too many students drop out of higher education because of a change of heart or because the course they chose did not meet their expectations, so do not rush into any decisions. When you have decided which subject or subjects you are interested in studying, you will then be faced with another set of choices. What kind of course do you want to follow? You could aim towards single honours in one subject or go for a course which leads to combined (joint) honours, covering at least two subjects. Then there are modular degrees to consider, on which you build up your individual course with short courses or modules selected from a much wider menu. There are also sandwich courses available on which you spend at least a year, or two periods of six months, in related work as part of the degree course. There are even combined subject courses on which you can enhance your study of science with, say, foreign languages, arts or business subjects.

A recent survey divided students into three groups: those who chose their degree as relevant to their employment prospects, those who chose it because they enjoyed the subject, and others whose motivation was less clear. 31% of the former went straight into jobs, 25% of the second group and 18% of the third. This does not necessarily mean, however, that the long-term job prospects of the latter were any the worse.

Sue Wood (page 44) put a lot of thought into her choice of degree course: *'In choosing university courses, I looked for a course with a high amount of practical work, a broad range of subjects in the first year and the ability to specialise in the last two years. I also wanted to go somewhere where the university and the department had a good reputation.'*

You will get lots of information from your school or college and through the careers service. There are many useful reference books such as *Which Degree?* and *Degree Course Guides*, as well as others in the Book List on page 163. You will discover that most degree courses require you to study other subjects at least for the first year, before you decide to go for single or joint honours. Single honours courses in sciences are available mainly in the older universities, whereas the newer universities and colleges usually offer combined or modular courses in science. Many universities offer a choice of single or combined (joint) honours courses and more variations are introduced every year.

Some recent developments in courses

- There are foundation courses and science degrees for people with A levels or equivalent qualifications in subjects other than science.

- Entry requirements are more flexible, e.g. A level and AS subjects, Advanced GNVQ/GSVQ, BTEC National, International Baccalaureate and, especially for Higher National Diploma courses, NVQ level 3.

- Longer four-year specialised courses in physics and engineering (accredited four-year courses are essential for chartered engineer status) are designed to be more compatible with European qualifications.

- More modular courses allow you to build up your own degree course with a series of short discrete modules covering subjects of your choice.

- A wider range of subject combinations is available, such as languages with sciences or business studies with sciences.

- New and unusual topics are on offer, such as equine science, digital imaging science, multimedia technology, health sciences for complementary medicine, environmental monitoring, mechatronics, to name just a few.

☐ More courses now include generic or key skills to improve graduate employability, and some work experience through links between universities and industry.

At some stage you will have to decide how broad or narrow you wish your course to be. Do you want to concentrate on one specialist area and build up more in-depth knowledge? This may be with a view to taking your interest further in a job, in research and development or in studying for a higher degree. You may, on the other hand, want a wider course which covers all aspects of, say, chemistry right up to finals or which includes at least two basic science subjects throughout. You might choose a course like materials science or environmental sciences which call on a variety of basic sciences in a multidisciplinary approach to a major topic. There are also some very specialist degree courses in subjects like minerals surveying, textile science, avionic systems engineering or prosthetics and orthotics, which lead towards very specific job areas.

Teacher training

There has been a shortage of science graduates - particularly in physics - entering teaching for some years, and recruits are currently offered a financial incentive or 'golden hello'. If you want to teach in state schools and train through a postgraduate certificate of education (PGCE), a substantial proportion of your first degree must be relevant to the National Curriculum. Check requirements carefully.

Jane Davies

Jane graduated in agricultural botany, which included enough biology for her to go on to train as a science teacher.

Career profile

Age: 30

Teacher of science at St John's School and Community College, Marlborough

A levels: biology, chemistry, maths (statistics)

BSc: Agricultural Botany at University of Wales, Aberystwyth

PGCE: University of Bath

'I always considered teaching as a possible career and had discussed the possibility during a school careers interview. However, I was very uncertain whether it would suit me, so I chose to maximise my career possibilities by selecting scientific A levels. I had always enjoyed biology and mathematics and wanted to study them further. I would have liked to have chosen history, but mixing humanities with science was difficult to do - so I chose chemistry as my third A level.

After my A levels, I decided to study a general scientific degree rather than complete a Bachelor of Education, as it would continue to keep my options open. I decided on a degree in Agricultural Botany, which was available at Reading University and the University of Wales, Aberystwyth. I chose Aberystwyth as it had a small, friendly department, and it worked very closely with the Welsh Plant Breeding Station at Trawsgoed, which offered valuable work experience opportunities during the vacations. It was also a campus university with a large percentage of accommodation on campus.

The course was excellent, as it included large amounts of genetics and biochemistry as well as botany. In my final year, I decided to train to be a science teacher specialising in biology. I applied to several teacher training colleges, and accepted an offer from Bath University. It was a one-year course, which involved two teaching practices in local schools.

After nine years, I still teach in the school where I got my first job. I have taught integrated science, which includes biology, chemistry and physics to GCSE level, and A level biology. In my first four years at the school, I was mainly involved in the organisation of the science department. I was responsible for year 7 science, which involved writing courses and coordinating staff and resources. I enjoy writing courses as you match the activities and practicals to the learning outcomes, and try to differentiate the work to all activities.

I have also been heavily involved in a science club during my time at the school, and the A level biology field trip to the Gower, which happens every September.

Given the career opportunities available in the school, I have, for the last four years, moved away from science administration and I am now the careers coordinator. I still teach about two-thirds of my timetable in the science department, but the remainder of my time is occupied by teaching personal and social education and careers-related work. I arrange careers interviews for the pupils in the school as well as writing the careers education component of the curriculum. I am also responsible for placing all year 11 pupils in work experience. This is particularly interesting as I have to find the places in the local community and then match the position to the needs of the pupil, to make it a positive experience for the company, the pupil and the school.

The role of careers coordinator has added a new dimension to my relationship with the pupils, as they no longer see me as just a science teacher and tutor. The teaching of personal and social education involves a completely different approach to teaching science, both of which I enjoy.

Teaching is extremely time-consuming but it is a very rewarding career. There are lots of opportunities to expand your experience and responsibilities, but, more importantly, the rewards of watching pupils achieve their goals makes it very worthwhile. It never becomes boring, as the courses and the expectations are continually evolving and each class responds differently to a given situation. There are always challenges for you as an individual.'

Incidentally, science graduates who particularly want to teach their own single science subject may find more opportunity to do so in the private education sector, as most comprehensive schools teach integrated science.

Combined degrees

Combined first degrees can be made up of complementary science subjects such as biochemistry and food science, or contrasting subjects such as science and business studies, chemistry and marketing, or science and languages. These contrasting combinations can often be useful for very specific careers where both subjects will be needed, such as technical marketing, export sales, purchasing and production engineering. The complementary combinations often form a wide general education as a basis for further study and training. Broad first degrees can be topped up with a specialist taught higher degree such as an MSc, although this takes a further year and financial support may be a problem.

Combined studies graduates have a wide range of career options. They follow similar paths to single honours graduates, although more go directly into work and fewer go on to further study. See chapter 7 for more details.

More or less technical: it's up to you

The direction you take will depend largely on your own interests and how they develop. If you are considering scientific research either in industry or as part of a higher degree, you would be advised to choose a fairly conventional single honours programme.

If you are sure you want your science degree to be part of a general scientific education - a base on which you can build either job training or further qualifications - you will not be disadvantaged by taking a broader first degree. In fact, in an area where you can use sciences in a particular work setting, such as teaching, information management or scientific journalism, it could really be an asset.

If you want to use your science degree in a completely unrelated field, such as one of the financial areas, it will not matter very much which way you go, although it is always an advantage to have done well, even if you don't want to go on with the academic subject. Employers are more impressed by someone with a good degree who wishes to change track, than by people who look as if they are abandoning their subject because they were not successful. Getting onto a postgraduate course, whether marketing, management or a specialist MSc, is also a lot easier if you have a good first degree. That usually means an upper second class or a first class degree. A good degree may also be a requirement for obtaining any funding at this level of study. So, again, following your interests and abilities can often be the best route to success.

Bear in mind that there are dangers in too narrow a specialisation. Your interests could take you into a very specific area where there may not be many jobs. On the other hand, you could suddenly become the most sought-after specialist on an unforeseen problem or development. You might strike it lucky, with employers falling over themselves to entice you to work for them. These situations are not easy to predict, but they do happen. For instance, the 1998 'New Frontiers in Science' exhibition included a breakthrough in the understanding of interaction of water molecules which could save billions of pounds a year through replacing oil-based materials with water-based ones. And research into the chemicals used by plants to communicate with insects may lead to increased crop yields.

Whole areas of science, such as biotechnology and polymer sciences, emerged and developed as breakthroughs were made in basic research. People worked on these areas long before they became part of the science courses in universities, so there is a constant reassessment of scientific knowledge and a demand for skills that is very hard to predict. Who knows what kinds of skill will be in demand in four or five years' time after you have completed your first degree, let alone two or more years later when you have finished a higher degree?

What do you want to do after you graduate?

Although many people begin higher education with some career plans, their ideas are likely to change as they find out more about themselves, their subjects and the career options open to them. Not surprisingly then, lots of people change their minds about their original choice of courses and about their career aims. A 1998 survey showed some 40% of students felt they had chosen the wrong field of study. Some students have a career in mind from when they are very young, but many others start out with very few ideas about what they want to do.

There are opportunities to move into business areas or other functions for those seeking to broaden their knowledge base as they progress. Alternatively you may prefer to develop as a specialist in your chosen field. Industry and society as a whole need both generalists and specialists.

Fortunately, most university and college courses are flexible enough to allow for change and development. But it does vary, so bear it in mind as a factor to look for when choosing a course. This flexibility will help to keep your options open and will give you the chance to experience some new areas before you are fully committed. When you consider possible courses, find out at what point you have to make your final decision on your degree subject, or subjects.

Paul Pilkington

Paul always wanted to be an engineer, so his choice of degree may not seem like the obvious one to make.

Career profile

Age: 24
Graduate trainee with British Aerospace Airbus
A levels: maths, physics and computing
BSc: Physics at University of Reading

'At school I was always fascinated by how things worked, I was always taking things apart and trying to put them back together again. My major interest has always been cars, so a career in engineering was the logical

choice. For GCSEs I was pointed towards maths, physics and computing with a foreign language (German) as a good basis for heading down the career path as an engineer. I also took geography, English and craft design (engineering).

At A level I was worried about what degree I needed to get into the motor industry and the response I was given was that I would need sponsorship from a company and high A level grades to get into the motor industry as a graduate. I was taking maths, physics and computing. This was supposedly the right combination.

In the end I chose to study for a degree in physics instead of engineering. It covered all the major topics of the engineering degrees, but in less detail, but also covered subjects like astrophysics and cosmology and particle physics. The degree course was enjoyable, as it was around 50% theory and 50% practical. From my point of view I was taking things apart to find out how they worked again!

I had applied for quite a few jobs covering electronics, vacuum systems, aerospace and television. I accepted a job at British Aerospace Airbus on the graduate training scheme. This gave me four placements in the first year and a single placement in the second year. In the first year I covered project management, test support (building new testing equipment) and fuel systems (for six months). My second year placement is still in fuel systems, which has given me a lot of experience. Even though I did not have an engineering degree, I have found that a lot of the work has on-the-job training. In addition, through the company's Education Resource Centre, I have taken part in a scheme which is run by the graduate employees themselves, going into schools to show pupils in practical ways how an aircraft is designed and manufactured.

In the long term, I am looking to take my career abroad to Toulouse in France, where I want to become a production support engineer for Airbus Aircraft on the production line, or perhaps move into customer support.'

Using your degree

The final year of your degree or diploma course is an important time to take stock and consider the options open to you. You can look again at your chosen science subject, or subjects, and decide whether or not you want to continue in more depth. Now is the time to explore all the different ways you can use your first degree in the next stage of your career.

There are three directions you can take:
- building on your scientific knowledge with further study, or research and development work in a related area
- adding to your scientific knowledge with a related vocational qualification or gaining work experience in an area where a science background is required
- using your degree subject and the other skills you have acquired in higher education to train in a completely different field.

Things to be aware of

- Nearly half of all vacancies for graduates are for people from any discipline - humanities, science, social science, etc. With a science degree, you could apply for any of these - plus some of the wide range of jobs which require a scientific background.
- In addition to your scientific knowledge, you can gain lots of skills from your undergraduate course that are transferable to any working situation.
- When you graduate, you will be looking for a first job: you will not be stuck there for the rest of your life. No new graduate wants to waste time and experience the frustration of a bad first job, but you can always move on. Even if you think you have made the right choice, you will probably change jobs as your career progresses.
- You can move from more scientific into less scientific areas much more easily than the other way round.
- Becoming too specialised means possible risks as well as rewards. It might lead to a dead end, or it could be the crock of gold at the end of the rainbow.

What you can offer employers

The science graduate who has more to offer than just scientific knowledge and skills is in a strong position. Employers want the kinds of transferable skills that young scientists learn during their training. This has been emphasised by both employers and careers advisers.

Transferable skills

The transferable skills of science graduates - most of which are similar to the Key Skills you will have gained at school - have been described as follows:

- organisational skills - proven by coping with a heavy workload of lectures, practicals and essay assignments; the ability to work to set deadlines, often several at a time, and the ability to prioritise work

- analytical and problem-solving skills - through both theory and practice, science undergraduates learn to ask pertinent questions, to interpret data critically and to deal with a number of variables simultaneously

Scientist at work by Daniel

- teamwork and communication skills - developed through working as a team in the laboratory, co-operating with colleagues but also working as an individual towards a team goal

- flexibility and adaptability - on science courses with a multidisciplinary approach, students have to take on board a wide range of scientific theories and techniques

- numeracy skills - with statistical knowledge, familiarity with mathematical software packages and experience of detailed data analysis, science graduates make numerate and computer-literate employees.

These are all skills which can be readily transferred to the workplace and can be added to your technical know-how. The articulate, numerate and, even better, literate science graduate, may have an advantage over non-scientists, even in competition for more generalist jobs. However, science graduates without these skills could be left behind unless they can compensate with really outstanding scientific ability and technical skills. A few very bright scientists work quietly away on their own, but they are a tiny minority. Proficiency in areas like organisation, teamwork and communication becomes essential for all young science graduates especially when the job market is tight.

Graduates from sandwich courses often move more quickly into first jobs than students from other courses. Their work experience, related to the subject of their course, is highly valued, and they are frequently offered their first jobs by their sandwich placement employer. The vocational approach of Higher National Diplomas is also appreciated by employers.

A degree of flexibility

It is the flexibility of a science degree that offers science graduates an advantage over other graduates in the hunt for their first job. The range of opportunities is clearly wider for those with science qualifications, since they can apply for all the general jobs and for more technical ones too. Science graduates can potentially be recruited to every activity within a large company. If science graduates have the additional skills and the motivation to secure these openings, they will clearly have an advantage.

Science graduates at the top?

The transferable skills of a science graduate as listed on page 92 sound like a recipe for a top job in industry, commerce, government or education, but do science degrees really take people to the top in Britain?

New graduates going into scientific, technical, engineering, research and development work in 1998 earned about £1000 more, on average, than trainees in management or accountancy and about the same as those in IT and computing - a median salary of £17,250. The median for all areas of graduate recruitment was £16,600. Some science graduates choosing to train as teachers receive cash incentives - and recruitment is beginning to pick up as a result.

The scientific community is more concerned about the low value placed on scientific expertise by industry and government in the past than about salaries alone. This situation is improving, and industry and the public sector do appreciate technical know-how. The shortage of people with scientific, technical and computing skills makes it more of a seller's market for these graduates. However, taking accountancy or management qualifications after graduating can still be a good way for scientists to gain promotion and financial rewards. This option attracts a steady stream of new science graduates and a trickle of postgraduates.

There is no doubt that starting salaries do affect the choice of a first job, but it is unwise to make the decision on that basis alone. The prospect of three or more years on a very low income may deter you from taking a PhD, but on the other hand you might be willing to put up with it for the freedom to work on your own project and your future prospects. There are lots of factors involved in choosing a career: money is just one of them.

One thing is certain: there will be changes ahead. There will be changes both in the way science graduates are used by employers and in what graduates themselves want from their careers. The development of information technology, the structure of European and international companies, and the growing recognition of the importance of a strong science base to the future of a successful British economy will all play a part.

Career development for science graduates will be discussed in more detail in chapter 9.

In conclusion

The dividing line between jobs for scientific specialists and more general positions is not absolutely clear-cut: it is possible to move from one to the other. But, as with choices of A level discussed in chapter 4, the decision made at the end of a degree or diploma course as to whether or not to go further with science is critical. Remember the one-way analogy: it is a lot easier to become less scientific in course and job progression than it is to become more so. It is easier to move from being a specialist scientist to being a generalist than it is the other way round (although Mike Partridge (page 118) is an example of somebody who has gone back into scientific research after starting a career in a more administrative post). People usually move away from the more mathematical towards the less mathematical. Many famous biologists have started as chemists or physicists, but it is very rare for mathematicians or physicists, famous or not, to begin as biologists.

A lot of people with science degrees progress into other jobs either soon after graduating or later on in their careers. In fact, several of the scientists profiled in this book have changed their areas of interest - such as from science into sales or management. Many of the communications and interpersonal skills acquired while doing a science degree are transferable into other areas of work. More and more employers are recognising this, and science graduates now realise that, despite the shortage of scientists and technicians, they must develop these transferable qualities if they are to compete successfully in today's job market.

People are sometimes put off scientific research because of fears that they might find themselves up a career cul-de-sac, but today's specialist skills could be tomorrow's major trend. It would make life a lot easier if we knew precisely which skills will be in demand in the future and where the opportunities will be located, but human resources planning is a very complex activity. While it is possible to make some general predictions, rapid changes in technology and a continuous stream of scientific discoveries make a detailed forecast very difficult indeed.

Careers with a Science Degree

Chapter 6: Where will your science degree lead?

This chapter:
- ☐ shows the context in which science graduates move on to the next stage
- ☐ details the first destinations of graduates from courses in maths, physics, chemistry, biological sciences and computing science
- ☐ compares these destinations with those of engineering graduates as well as graduates as a whole
- ☐ provides examples of first jobs taken by science graduates
- ☐ discusses ways in which you can use this information.

The background

The most quoted destination statistics tell us how many graduates go into full-time jobs; in fact the new Government 'employability' league tables are based on just that. What they do not tell us is how many graduates go on to postgraduate study, teaching and other professional training, take temporary jobs or are overseas students returning home.

All universities and colleges offering degree courses have careers services to help students with their career choices and with finding jobs or further courses. It is these services that collect data regarding what students are doing six months after graduation. This information is amalgamated with that from other universities and colleges to produce national data. A digest of this is published every year in *What Do Graduates Do?*

First employment figures don't tell you:
- ☐ what the employment situation is like for all graduates; not everyone responds to the survey
- ☐ where new graduates go after this information is collected
- ☐ what the equivalent unemployment rates are for those without a degree
- ☐ what happens to those who go on to further study or training.

So is it worth doing a degree?

With newspapers full of headlines about student debt, you may be wondering if it is worth doing a degree at all. A science course is hard work and a degree is a substantial investment of time and money. Your family may have to make sacrifices to support your higher education and there is an expectation that this should 'pay off'. Many parents - and students - feel that the knowledge gained on a degree course should be of immediate use in a job and that being a graduate will give you instant status and enhanced opportunities.

On the whole, higher qualifications do pay off, but not necessarily straight away. Statistics show that, by their early thirties, graduates are earning between 12% and 38% more than their non-graduate contemporaries who have at least one A level. The pay gap is greater for women than for men. A degree is a starting point for many rewarding and satisfying careers, but your ultimate success may well depend as much on your personal qualities as on the certificate you receive at the graduation ceremony.

However, the class of your degree does make a difference to your chances of being employed six months after graduating. Graduates with a first-class honours degree are considerably less likely to be unemployed than of those with a second-class degree, and so on down the scale. A higher proportion of graduates with first class honours stay on for further study and training, so there will be proportionately fewer looking for jobs.

Although the rise and fall of Britain's economic fortunes affects new graduates as it affects everyone, graduate careers services report that they are usually the last to be hit by a recession and the first to benefit from recovery. In 1995, people with no qualifications were four times more likely to be unemployed than graduates.

All the evidence shows a continued commitment on the part of employers to the long-term recruitment of graduates, and a growing realisation of the need for a better educated workforce if we are to keep up with our industrial competitors at home and abroad.

Stand up and be counted!

One of the biggest changes in the employment market for new graduates is that there are a lot more of them than there used to be. Not only are there many more universities producing graduates, but there are also a large number of colleges of higher education. Altogether these institutions now turn out well over 200,000 graduates per year. In addition, there are over 15,000 students a year leaving colleges and universities with Higher National Diplomas. And we have still not counted all the people studying for degrees part time, including Open University students! Apart from those going on to take further full-time courses, most of these people are coming onto the job market at about the same time each year.

At the start of the 1980s one in seven people from the 18 to 19-year-old age group went into higher education; now it is nearer one in three, and there are a lot of people starting degrees later in life as 'mature students'. So graduates don't have the rarity value they had when many of your parents were in their early twenties.

But the increase in science, technology and engineering graduates has not matched the increase in graduates generally. The advantage for the scarcer science graduate is that the competition for jobs which require a science or an engineering degree has not risen as it has for positions open to graduates of any discipline. In fact, there is a shortage of qualified people.

Whereas about a fifth of graduates go on to further study and training, a higher proportion of science graduates do so; over 40% of 1997 chemistry graduates, for example.

What is a graduate job?

There is much discussion about what constitutes a graduate job and distinguishes it from any other job. There are some professions, such as medicine, pharmacy or veterinary science, where it is essential to have taken the relevant degree courses in order to do the job. However, there are many other areas of work where the rules are far less clear-cut. Many jobs now taken by graduates would have been done in the past by people with A levels and some even by those leaving school at 16. A recent survey showed that 30% of graduates are in a job that does not require a degree, but the proportion of jobs calling for a degree rose from 9.7% to 14.1% between 1986 and 1997.

Young graduates are attractive to many employers because they do not need as much training and require less supervision than school-leavers (although there are lots of unfavourable comments in the press about lack of literacy and other transferable or Key

Skills!). The jobs themselves have changed too. There are science graduates now taking lab technician jobs in the place of school-leavers who would in the past have trained on the job. This is an inevitable consequence of more people going on to higher education: employers looking for good candidates now take graduates because people who might, in the past, have left school at 18 are now doing degrees.

There are many reasons why graduates might initially take such jobs: no opportunities in the area in which they really want to work, accumulated debts from student days, uncertainty about what they want to do and so on. But many graduates take jobs which do not require a degree just as a temporary measure, before the right job turns up. Even an 'unsuitable' job can put you in the right place at the right time, if you take every opportunity to prove your potential for better things.

Not all science graduates are cut out for a career in scientific research, but there are many other types of work where a science degree would be useful and could lead to a more fulfilling job.

There is now a much wider range of graduates - more women graduates and more mature graduates. Graduates themselves are choosing routes and lifestyles that are very different from the traditional roles of the past. Employment forecasts predict a continuing growth in jobs for graduates in general and science and technology graduates in particular. There will be a corresponding continued decline in jobs for people with few skills and qualifications.

So it might appear that a graduate job is best defined as any job a graduate is prepared to do. We do hear stories of the graduate milkman, the PhD postman, etc, but remember, these cases are newsworthy because they are not the norm. Someone who has worked for three or more years to pass academic exams can and should expect a rewarding career that reflects their academic achievements.

On graduation, there is an expectation of a new beginning and of great opportunities opening up. But it is a mistake to expect that you will have an automatic right to a 'good job' or a meal ticket for life just because you are a graduate. Getting the right job for you depends not only on the opportunities available but on all sorts of other factors such as your personal qualities, the skills you have, where you live, where you want to work, what your career commitment is, what kind of lifestyle you want, your health, luck and so on.

Why do employers choose graduates?

- Graduates are usually employed for their future potential rather than for any knowledge or experience they bring to their first jobs.

- Graduates are expected to be 'self-starters' who can manage their own work. This means they are more independent in their posts in a way that school-leavers are not, and are able to put their own stamp on the job.

- As the working environment becomes even more complex, with new technology and legal requirements, new industries replacing old ones, changing business practices, more advanced products, wider markets etc, so jobs become increasingly involved and require well-trained and highly educated people to cope with them.

How do new graduates find jobs?

The traditional graduate route into employment was to job hunt in the final year of the course when employers visited the universities on the so-called 'milk round', when recruiters travelled the country demonstrating their business and interviewing candidates on the spot. Students would apply to large companies, through their graduate recruitment schemes, for traineeships leading to one, two or more years of training. Science graduates going into research and development, however, were usually recruited directly into a job.

Employers now recruit graduates in many different ways: the graduate traineeship route has declined in importance and involves only a small proportion of new graduates today. Press advertising is the most popular method of recruitment, and most graduates now enter employment by applying for specific jobs or by taking temporary project work on short-term contracts. They may still make their way up through an organisation by performing well in each job, but they are now more likely to have to use their experience to move elsewhere. An increasing number become self-employed. This is very different from the situation when your parents' generation left university.

Much jobsearching and applying is now done 'on-line' through the Internet. It is possible to send your CV, instantly, to any employers you are interested in, and also to find out about the companies and any vacancies they may have. There has also been a growth in telephone and Internet interviewing - at least as a preliminary measure.

Another change is that small and medium-size businesses are now major recruiters of graduates; it's no longer just the domain of the 'blue chip' companies. Manufacturing and industrial services still provide 40% of graduate vacancies, but the non-industrial sector is more optimistic about recruitment in the next few years than the industrial sector.

Many employers prefer to recruit graduates with one or two years' experience, rather than graduates fresh from university or college. By selecting older applicants, employers recruit workers with higher levels of maturity and personal and communication skills, together with the ability to perform effectively from day one. Older recruits tend to stay put with an employer for longer too. This means that for many new graduates the first job can be seen as a stepping stone to better prospects a year or two ahead.

First destinations

Not all graduates look for work immediately after finishing their courses. Many science graduates go on to higher degrees or take training courses for specific jobs such as teaching. Others may not know what they want to do. They may want to take stock of things; do voluntary work, short-term unskilled jobs or spend a few months travelling. Many of these people will be counted as unemployed in the statistics collected at the end of December, but 12 months later they could well be on training courses or in permanent jobs. Today there are so many routes from higher education into work that the first destination figures can only give part of the picture.

Students who are on sandwich courses in science and technology are among the quickest to find jobs, often with an employer they have worked for as part of their industrial placement.

The first of the following tables shows the first destination by category of graduates completing single subject degrees in maths, physics, chemistry, biology and environmental sciences in 1997. The second table gives first destinations of all graduates, computing science graduates and, for comparison, those completing degrees in electronic engineering and mechanical engineering.

Chapter 6 - Where will your science degree lead?

Key

A = in UK employment
B = in overseas employment
C = studying in the UK for a higher degree
D = studying in the UK for a diploma, certificate or professional qualification in teaching
E = undertaking other further study or training in the UK
F = undertaking further study or training overseas
G = not available for further study or training
H = believed to be unemployed
I = seeking employment, study or training but not unemployed.

	Maths	Physics	Chemistry	Biology	Environmental sciences
A	59%	49%	48%	53%	63%
B	1%	2%	2%	2%	2%
C	13%	31%	33%	18%	10%
D	10%	4%	5%	6%	4%
E	5%	2%	3%	4%	3%
F	*	*	1%	1%	1%
G	4%	4%	4%	6%	5%
H	6%	6%	5%	9%	11%
I	1%	1%	1%	2%	2%

	Computing science	Electronic engineering	Mechanical engineering	All graduates
A	82%	75%	74%	65%
B	2%	2%	2%	3%
C	5%	9%	10%	9%
D	1%	1%	1%	4%
E	1%	2%	2%	6%
F	*	*	*	*
G	2%	2%	4%	5%
H	6%	9%	6%	7%
I	1%	1%	1%	1%

Statistics from 'What do Graduates Do?' 1999

* = less than 0.5%.

Careers with a Science Degree

Points to note about these tables

- ☐ The figures are rounded up or down, so may not add up to 100%.
- ☐ Medical and veterinary graduates are excluded, as they go on to clinical training.
- ☐ HND holders are not included, but 57% of full-time course and 44% of sandwich-course completers went on to study for a first degree in the UK; it is possible to go on to do a higher degree but this is very unusual unless the HND holder also has considerable experience to offer.
- ☐ At the time the unemployment rate for the whole 21-24 age group was 11.5%.
- ☐ Physics and chemistry graduates are more likely to continue on to further study than other graduates.
- ☐ Fewer engineering graduates than science graduates take higher degrees; they need relevant working experience after an accredited first degree to gain a professional qualification.

How are science graduates employed?

Below are some examples of first jobs entered in 1997 by a sample of new graduates with first degrees in science subjects.

These are only a few examples to give an idea of the range of jobs graduates go into. It is by no means an exhaustive list.

Scientist at work by Jade

Biology graduates

- ☐ Sales adviser for an electricity company
- ☐ Events administrator

- [ ] Production manager in biotechnology
- [ ] Water purification biologist
- [ ] Laboratory technician
- [ ] University research technician
- [ ] DNA analyst
- [ ] Accountancy trainee
- [ ] Clinical trials monitor
- [ ] Field trials officer
- [ ] Research scientist
- [ ] Trainee systems programmer
- [ ] Marketing trainee
- [ ] Salesperson
- [ ] Air cabin crew
- [ ] Medical salespersons
- [ ] Police officer

Chemistry graduates

- [ ] Banking and insurance managers
- [ ] Army officer
- [ ] Analytical and research chemists in fuel, pharmaceutical, food and other industries
- [ ] Research and development scientists
- [ ] Teachers
- [ ] Accountants
- [ ] Computer analysts and programmers
- [ ] Environmental consultant
- [ ] Product development scientist
- [ ] University research assistants
- [ ] Food technologist

Computing science graduates

- [ ] Computer system and data processing managers
- [ ] Banking and insurance managers and administrators

- ☐ Computer software engineers and technicians
- ☐ Computer analysts and programmers in industry, commerce, the media and the public sector
- ☐ Software consultants
- ☐ Research scientists
- ☐ Design engineers
- ☐ Development engineers

Electronic engineering graduates

- ☐ Management trainees
- ☐ Communications engineers
- ☐ Electronics engineers
- ☐ Software engineers
- ☐ Design and development engineers
- ☐ Process and production engineers
- ☐ Systems engineers
- ☐ Test engineers
- ☐ Computer analysts and programmers
- ☐ Marketing officer
- ☐ RAF pilot

Environmental sciences graduates

- ☐ Managers and administrators in finance, industry and Civil Service agencies
- ☐ Local authority development officer
- ☐ Wildlife Trust assistant reserve officer
- ☐ Retail manager
- ☐ Environmental consultants
- ☐ Environmental control officers
- ☐ Environmental planners
- ☐ Conservation workers
- ☐ Environmental scientists
- ☐ Field surveyors

- ☐ Geographical information systems assistants
- ☐ IT specialists
- ☐ Mapping and charting officer
- ☐ University research assistants
- ☐ Systems engineer
- ☐ Water quality analyst
- ☐ Police officer

Mathematics graduates
- ☐ Marketing and sales managers and analysts
- ☐ Financial management trainees
- ☐ Software engineers
- ☐ Accountants
- ☐ Statisticians
- ☐ Actuarial trainees
- ☐ Computer analysts, programmers and modellers
- ☐ Systems engineers
- ☐ Stockbrokers
- ☐ Investment analysts
- ☐ Insurance underwriters
- ☐ Scientific researchers
- ☐ Website editor
- ☐ Civil servants
- ☐ Stock assistants

Mechanical engineering graduates
- ☐ Officers in the Armed Services
- ☐ Managers and administrators
- ☐ Mechanical engineers in fuel, marine, automobile, mining, aeronautical and other fields of engineering
- ☐ Design engineers
- ☐ Research and development engineers
- ☐ Process and production engineers

> Careers with a Science Degree

- Software engineers
- Computer analysts and programmers
- Service engineer

Physics graduates

- Officers in the Armed Services
- Managers and administrators
- Production manager
- Reactor physicist
- Research scientists
- Software engineers
- Design and development engineers
- Process engineer
- Systems engineer
- Test engineer
- Accountants
- Computer analysts and programmers
- Actuarial trainee
- Investment and merchant bankers
- Pensions adviser
- Statistician

Source: Information taken from *What Do Graduates Do?* 1999.

Points to note about these lists

- They do not include clerical and secretarial posts, sales assistants etc, of which there are a number in all categories.
- The proportion of graduates going into employment related to their subject varies; computing science can claim about two thirds, while less than one third of biology and physics graduates enter subject-related jobs.
- All subjects include people entering employment in management, administration and computing.
- Graduates are significantly less likely to be unemployed than the rest of the population.

Chapter 6 - Where will your science degree lead?

- Shortages of graduates are forecast in computing, electrical and electronic engineering, food science and chemistry.
- Growth areas in science and technology are the pharmaceutical industry and biotechnology.

Paul Compton

Paul chose a degree course specifically designed for his chosen career, and is well aware that the learning process does not stop once you have graduated.

Career Profile
Age: 27
Optometrist, working within the Haine and Smith partnership.
GCSEs: mathematics, biology, chemistry, physics, English (lang. and lit.), art, technical design and drawing, French, music.
AS level: mathematics.
A levels: mathematics, biology, chemistry.
BSc: Ophthalmic Optics (Hons), Cardiff University.

'My choice of GCSEs reflected my interest and enjoyment of science-related subjects, although at this time I was unsure if I wanted to pursue a science career or a career in design and architecture. Art and technical design were necessary to keep options open, French was a compulsory foreign language and music I simply enjoyed. My GCSE results revealed superior grades in the sciences, which convinced me that I should seriously consider a science degree.

For A level I chose biology, chemistry and mathematics. My weakest subject of the three was mathematics and it was decided that I should take AS level at the end of my first year of sixth form followed by A level mathematics in the second year of sixth form.

One area of biology I particularly enjoyed was anatomy and physiology, therefore I began to investigate degrees related to medical science and research.

A period of work experience with a local optometrist gave me valuable insight into a career which previously I had known very little about. I applied to the five universities offering degrees in Ophthalmic Optics and decided to make Cardiff University my first choice.

The ophthalmic optics degree was a three-year degree course consisting of lectures, practicals and actual clinical work. The first year of lectures ensured that I understood the basics of biology, chemistry and physics in relation to the human body and in particular the eye. The lectures were complemented by series of practicals in optics and physiology as well as sessions in the department's eye clinic, to develop clinical skills in using some of the equipment used during an eye examination.

Second year lectures covered more in-depth physiology of the human head, brain and eye, as well as optics and other clinical subjects. Physiology practicals continued, and in the eye clinic I was prepared to examine my first patients under supervision at the end of the year. Contact lens fitting on fellow optometry students was also carried out.

The third year devoted more time to clinical studies, and in the eye clinic I routinely examined eyes and fitted contact lenses, still under close supervision. A project is given to each student in their final year. I was to carry out research on a new test being developed for use when examining children. This gave me invaluable experience in testing children, as well as giving me an insight into research and the skills that would be required for further education.

Attaining a degree in ophthalmic optics was the first stage in a two-part process to become registered as an optometrist. The second stage was to work in private practice as a pre-registration optometrist and to take the professional qualifying examinations of the College of Optometrists. I joined Andrew Matheson Optometrists as a pre-registration optometrist, working under supervision of a qualified optometrist. University had prepared me well for the theoretical aspects of optometry, but only time in practice dealing with patients helped to develop communication skills as well as practical knowledge in dealing with patients. During this period I was able to spend half a day a week in the eye department of the local hospital, where I was to gain invaluable experience in assessing unusual eye conditions as well as insight into the role of the hospital eye service.

On passing the professional qualifying examinations, I was able to register as a qualified optometrist with the General Optical Council and was now free to work unsupervised. I remained with Andrew Matheson Optometrists until June 1995 when I moved to Wiltshire and joined the Haine and Smith partnership working as an optometrist.

Haine and Smith Opticians is a Wiltshire-based firm consisting of nineteen practices In my role as optometrist, I travel around five practices carrying out eye examinations and fitting contact lenses. Within each practice I work as part of a team, alongside optical assistants and dispensing opticians. Being an optometrist provides many challenges on a daily basis, and there is great job satisfaction in providing a service which not only helps patients *make the most of their vision, but also through careful examination can reveal other problems, both ocular and systemic, which need further investigation by a general practitioner or other specialist. As well as an interest in pathological conditions, I also enjoy the challenge of fitting contact lenses, which can give patients more freedom to be able to take part in activities such as sport, which previously would have been restricted when wearing glasses. My role within Haine and Smith has developed, and I have recently joined the contact lens board. This new role combines my work as an optometrist along with an input into the business aspect of contact lenses.*

My plans for the future are to stay with Haine and Smith and to develop my role within the partnership. I would like to expand my expertise and skills by taking higher qualifications. Opportunities exist to take certificates and diplomas. I am particularly interested in continuing my education in contact lens practice. Opportunities also exist for MSc and other higher qualifications. At present, compulsory continuing education is being debated within the profession and I strongly agree on the need to continually update knowledge.

I believe as an optometrist that I am well paid. Those who are interested in a career in optometry need good communication skills, an ability to analyse a series of signs and symptoms and to record information accurately.'

Career development for science graduates

This chapter has covered the activities of science graduates for the six months after graduating. As we have observed, your first post is unlikely to be a job for life. What happens after this period is much harder to determine. Many people will move on to other jobs in two or three years' time. Career development will be covered in detail in chapter 9, after the information about further courses and international opportunities for science graduates.

Careers with a Science Degree

Chapter 7: Postgraduate study - what are the options?

This chapter:
- outlines the main options for further academic study and training
- looks at some examples of postgraduate courses taken by science graduates.

Why study after a degree?

Each year, nearly 20% of first degree graduates go on to do some kind of postgraduate study or training immediately after their degree course, and many more will do so later on after some working experience.

There are several reasons why
- For career development
  - to develop knowledge of your own field
  - to gain knowledge of a specialised field
  - to change fields
  - to improve your career prospects in a difficult job market
- For scholarship and personal challenge
- To give yourself more time to decide on your future aims.

Before you make any firm decisions, it is important to ask yourself:
- will there be a demand for the kind of postgraduate qualification I am embarking on?
- do I still have the motivation and interest to pursue further study?
- do I have the funds to continue with further study?

Those with first class honours degrees and upper seconds are more likely to go on to further academic study because they have a better chance of getting financial support or studentships.

Careers with a Science Degree

Percentages of graduates going on to further study or training 1997

	Further academic study	Teacher training	Other training
All graduates	8	4	6.4
Biology	18	6	3.8
Chemistry	32.7	4.6	2.6
Computer Science	5	1	1
Maths	13.2	10.3	4.5
Mech. Engineering	10	0.8	1.8
Physics	31.4	4.2	2.2

Source: Statistical information taken from What Do Graduates Do? 1999

Points to note about this table

☐ More science graduates than other graduates go on to further study or training at the end of their degree courses.

☐ Engineering graduates are likely to go on to approved work experience towards a professional qualification rather than academic study or training.

☐ Comparison with previous years shows that there has been a steady increase in the proportion of graduates going straight on for further qualifications after their first degrees. This may be due in part to difficulties in the employment market, but also reflects a continual increase in demand for qualifications.

☐ The table covers only those going on to full-time further study and not the many others who do part-time further study in their first job or later on. Many employers of scientists encourage them to do higher degrees part-time. Some science graduates decide on their own to return to education full-time for very similar reasons to those given above.

The three types of postgraduate qualification

- Higher degrees by research
- Higher degrees by instruction
- Vocational courses leading to certificates or diplomas, or to professional qualifications such as teaching, accountancy, law and social work

All of these qualifications can be full-time or part-time, although higher degrees by research are usually full-time.

There is no state funding support to follow the full-time route except for postgraduate teacher training, so you will need to find a grant or a studentship or be self-financing. Part-time courses are often done while working, with some financial support from the employer.

There are over 6000 postgraduate courses in the UK - some on very specialist topics. So when course titles are mentioned in this chapter it will probably be only one of a number in the same area of work. Only a selected few are mentioned to give a flavour of the range of options available. There are reference books with comprehensive lists of all the current courses which you would consult in your final year on a degree or diploma course (see Book List, page 163).

Higher degrees by research

Higher degrees by research lead to a doctorate (a PhD or DPhil, depending on what it is called by the university awarding it) or an MSc (Master of Science). A PhD takes a minimum of three years and an MSc usually takes 12 months' full-time or two years' part-time study. After a PhD you can use the title Doctor, although you could find yourself having to explain in a medical emergency that you are not a 'proper' doctor!

Higher degrees by research are usually taken in a university, research institute or industrial laboratory which has a link with a university. Some universities require all research students to take an MSc first and then, if this goes well and finance is available, they go on to a PhD afterwards. For more information about the research degrees on offer, all departments have been given a research rating by the Higher Education Funding Council which reflects the quality of the research being carried out, and the publications issued.

Learning to do research

A degree by research involves an in-depth study of a very specific area. Each research student has an academic supervisor who acts as a guide and mentor. The results of the research are presented in a thesis - a detailed written discussion of your project - which must include work that is original and makes a contribution to the understanding of the field.

A higher degree of this type is really a training programme in research, and involves:

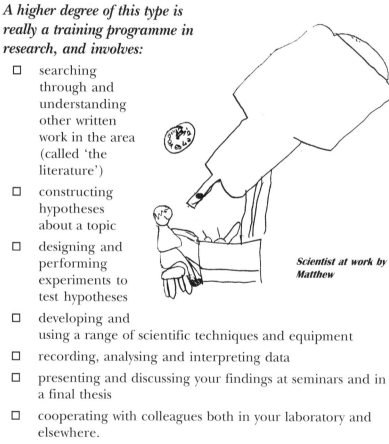

Scientist at work by Matthew

- searching through and understanding other written work in the area (called 'the literature')
- constructing hypotheses about a topic
- designing and performing experiments to test hypotheses
- developing and using a range of scientific techniques and equipment
- recording, analysing and interpreting data
- presenting and discussing your findings at seminars and in a final thesis
- cooperating with colleagues both in your laboratory and elsewhere.

Postgraduate research is often seen as a solitary occupation, with students working mainly alone in libraries. This is not true for scientific and technical research where you would be based in a laboratory, interacting on a daily basis with other scientists in your own institute and maybe with scientists in labs around the world, through conferences and computer link-ups.

Nonetheless, what counts for your PhD is your own laboratory work. You will need persistence and determination to learn the necessary techniques and complete exacting and often repetitive practical work. Most first degree courses include some project work in addition to laboratory practical classes, and this experience will give you some idea of whether you would like research work. Although the chosen topic can be important in terms of further career progression, it often turns out that people who make a career of science find themselves working on very different subjects later on.

A higher degree by research is not only a training for academic research; many people with PhD and MSc qualifications also work in industry, teaching, administration and other areas. Employment information shows a very low rate of unemployment among people with PhDs.

Ten years after their doctorates or Master's degrees many senior scientists do very little laboratory-based work. They could be supervising other scientists or may have moved into other areas of work where the training and experience of research is of great benefit.

Funding research degrees

Research students get funding from a variety of different sources.

There are studentships funded through government grants distributed through various research councils, through science-based trusts and foundations such as the Wellcome Trust and the Nuffield Foundation and through industrial sponsorship. The funding has to cover living expenses, research materials and academic fees. A list of addresses of research councils can be found on page 165.

The amount provided for living expenses varies depending on the source, but it is considerably less than you would get if you went into a job. People who get support from research foundations or employers are better off than those on the basic postgraduate studentships. The average funding from research foundations is approximately £7000 in London, and £5500 elsewhere.

There are also Co-operative Awards in Science and Engineering (CASE) which support research students working on projects jointly devised and supervised by academic departments and partners in industrial research labs. The Particle Physics and Astronomy Research Council has launched CASE Plus which offers an additional year where the student works full-time on the premises of the firm and receives a salary of up to £24,000.

Another option is to take out a loan. Career Development Loans are operated on behalf of the government by certain banks. During the course and for a month afterwards (up to 18 months if you remain unemployed) the government pays the interest on the loan.

Scholarships are offered both by universities and independent bodies. They are advertised through academic departments, scholarship offices, careers services and the education and specialist press.

Investigate the possibility of European funding, as this is available particularly in science and engineering.

If you fail to find funding, you could think about taking a paid research position. Research and Teaching Assistantships are advertised in education and specialist press, e.g. the Guardian, THES, New Scientist. This is not an easy option as you have to manage a large teaching load alongside your own research.

A research student has to regard this period as an investment in future career prospects. Competition for funding is keen and it is very difficult to get any at all unless you have either a first class or upper second degree.

Mike Partridge

In this profile, you will read about a physics graduate who now works as a research fellow for the Institute of Cancer Research. Previously, as a scientific administrator with the Engineering and Physical Sciences Research Council, he was involved with reviewing grant applications and working out funding priorities.

Career Profile
Age: 29
Research Fellow
Institute of Cancer Research, Sutton
A-levels: physics, chemistry, mathematics, further mathematics.
Degree: Natural Sciences, The University of Cambridge
PhD: physics, Cranfield University

'I left sixth form college with a good selection of science and maths A levels, but did not have any clear career ambitions. I chose the Natural Sciences

Chapter 7 - Postgraduate study - what are the options?

course at Cambridge University because it allowed me to study a broad range of subjects over the first two years, and only specialise towards the end of the course. I studied the core of the physics course but was also able to branch out into some new subjects, spending my final year studying material sciences and metallurgy. The practical yet pragmatic approach of this course, combining theory, mathematical modelling and experimental work was something I really enjoyed and has shaped the way I have worked ever since.

After my first degree, I was keen to continue working in science, so took a job as a research scientist with Cranfield University. The job was a small defence research contract investigating computer night vision systems, so involved writing regular reports for the sponsor and visiting them about once a month. While working at Cranfield, I registered as a part-time PhD student and was able to write up the work I was doing on the industrial contract in my spare time. I got great satisfaction from this work and really enjoyed the opportunity to spent time studying a small field in great depth, and develop my skills as a scientist.

After four years at Cranfield University, with funding from research contracts increasingly difficult to find, I left and took up a post as a scientific administrator at EPSRC (the Engineering and Physical Sciences Research Council). EPSRC is a government body that is responsible for allocating public money for scientific research in universities within the UK. My job there involved two main tasks: operating a peer review system for grant applications, and working out future funding priorities. The peer review system involves sending each request for funding received to several different leading scientists in a relevant field for their opinions about how important it was; this was largely just an administrative task. The second part of the job involved visiting government and industry to find out where research was needed and then visiting universities to see if they were in a position to meet these needs. EPSRC was then able to target funding to places where money was most needed to solve current and future problems in the UK. The programmes I worked on particularly were in recycling, reduction of pollution and renovation of contaminated land as well as research into tunnelling, road construction and coastal and flood defences.

Although I found this job very interesting, I felt that what I had really enjoyed more than anything else was being a scientist. After one year at EPSRC, I left and took up a research fellowship with the Institute of Cancer Research at the Royal Marsden Hospital. My work there involves research into radiotherapy treatment for cancer, especially the use of computerised x-ray imaging systems. I find my work at the Institute academically stimulating and personally very rewarding. I work alongside clinical staff, providing technical support for clinical trials, and research new methods of treatment.

My career so far has taken me to several very different fields, from defence to environmental civil engineering to healthcare. Each time I moved, it took time to learn the new challenges and peculiarities of each field, but the core skills of a research physicist remain the same, and skills learned in one field were easily transferred to another. I am very happy in my current post and hope to continue researching in this field for the foreseeable future.'

Jessica Harris

Jessica works as a medical statistician, and received funding for her postgraduate course from the Medical Research Council.

Career Profile

Age 30

Medical Statistician, National Heart & Lung Institute, Imperial College

A levels: Mathematics, Economics, Sociology

BSc: Mathematical Statistics, University of Liverpool

MSc: Medical Statistics, London School of Hygiene and Tropical Medicine, University of London

'I enjoyed maths at school, although I found statistics easier to understand than the pure maths component of the A level. Originally I intended to follow a career in economics, and applied to do a degree at Liverpool University which would allow me to study both statistics and economics. The economics part of my degree course was introduced in the final year, but by this time I had decided to complete my degree only in statistics as I felt it would be more sensible to specialise in one subject. In my final year, I took a medical statistics course, using statistical techniques within health research, which I enjoyed, and my lecturer encouraged me to apply to the MSc course in London.

The MSc course was very different from my first degree, very applied and vocational. It took a year, and the course director in London helped me get funding from the Medical Research Council to cover my fees and living expenses in London. Broadly speaking, graduates in medical statistics tend to work in academia, or for pharmaceutical companies. I decided academia would suit me better.

There always seems to be demand for medical statisticians, so I didn't find it too difficult to find a job. I started working at the National Heart and Lung Institute straight after graduating and have stayed in the same post.

In our department we undertake a number of studies to try to understand what environmental factors cause asthma, and other lung diseases. In general, most studies involve collecting information on a group of individuals, and looking at the relationships between health and the environment within these. Some of the studies take a long time. For instance, in one study, we have followed a group of children since before birth to try to determine what causes asthma - this has been running for five years. Other studies include occupational surveys, where we are invited by companies where workers may be exposed to possible triggers for asthma to visit the site and do a health survey on all employees. This would involve administering questionnaires and allergy testing all the employees.

As the senior statistician, I am involved in the design of studies, data management and all statistical analyses of the projects I work on. I am also responsible for contributing towards reporting results, in medical journals (such as the British Medical Journal) and at scientific meetings. For the health surveys, I am often involved in all aspects of these studies, including some of the fieldwork, which means my job is quite varied.

I find the work challenging and interesting. There is scope for collaborative work outside the department; we have a group of colleagues in Europe who have common aims whom we often work with. I am often asked for advice and statistical assistance within the Institute or our associated hospital, Royal Brompton Hospital, which has given me an opportunity to work in different areas of medicine. I am invited to give statistics lectures frequently, and ran a workshop last year.

There have also been opportunities to continue my education. A couple of years ago I registered for a part-time PhD in epidemiology and I am also registered for a postgraduate diploma (also in epidemiology) at McGill University in Montreal, Canada. Epidemiology is the study of the distribution and causes of illness in populations, which is the methodology behind this type of medical research and goes hand-in-hand with medical statistics.

Being able to use statistical techniques in this environment has suited me very well; friends from university have gone into finance and marketing, but I have always had an interest in medicine and the work feels 'worthwhile'. There seems to be increasing demand for suitably-trained medical statisticians, many offering work abroad in developing countries which I would like to try in the future. In the long term, I would like to complete my PhD and hopefully get a lectureship post.'

Higher degrees by instruction

Taught MSc courses

Here are a few examples of taught MSc (Master's) courses: (course requirements are given in brackets)

- ☐ Computer systems design (degree with significant computer science content)
- ☐ Control engineering (graduates of any discipline)
- ☐ Hydrogeology (science degree plus maths beyond A-level)
- ☐ Fish biology and management (biological sciences degree)
- ☐ Optoelectronics and laser devices (electronics or physics degree)
- ☐ Occupational safety and health (any degree)
- ☐ Colour chemistry (science or chemical engineering degree)
- ☐ Environmental chemistry (chemistry degree)
- ☐ Statistics (variety of different courses requiring different degree backgrounds)
- ☐ Operational research (degree with high mathematics content)

The growth in scientific and technical knowledge is too extensive for so many specialist topics to be absorbed into first degree courses, making postgraduate taught courses necessary to cover areas like these.

As you can see, these courses recruit from a wide variety of degree backgrounds. A particular subject at first degree level may be required, but this is not always the case. Some MSc courses also prefer people to have appropriate working experience. Master's degrees in areas like information science and clinical psychology involve periods of relevant practical experience and lead to professional qualifications.

An MSc usually lasts a complete calendar year full-time or two years of part-time study. Master's courses include coursework, lectures and seminars, together with practical work and training in research methods. They usually end with a combination of final examinations and a short thesis.

For some science graduates an MSc is an introduction to a new area of work such as law or social work, while for others, particularly those with a combined studies degree, it may involve a more specialist study of topics touched on in their first degrees. Some MSc courses are arranged by employers in collaboration with a university as part of their graduate development programme.

Whatever your reasons for considering further training, you will need to take stock of your situation and do some career planning. Before committing yourself to any Master's course you need to check out the employment prospects carefully. Even though some course titles sound very vocational, there may not be many related job opportunities. Some postgraduate courses recruit a lot of students from overseas, for example, because their employment prospects in their own countries will be enhanced. This does not necessarily mean that home students seeking work in the UK will have the same advantage. However, because a full-time MSc lasts only one year, the job situation should not change too much from the start of the course to the end. The experience of recent leavers will help you to predict your chances.

MBA

One very popular Master's course is the Master of Business Administration (MBA). This is taken by many scientists, even those with PhDs, because they want to enhance their opportunities in management. People are strongly advised to gain some working experience before embarking on an MBA course: most science graduates taking an MBA are several years beyond graduation and will have managed other scientists or research teams.

PGCE

The best known postgraduate certificate course is probably the Postgraduate Certificate in Education (PGCE). This qualification is required by graduates wishing to teach in state nursery, primary or secondary schools. The proportion of science graduates taking PGCEs is shown in the table on page 114.

Science graduates who apply for a teacher-training course in science need to have taken basic sciences as a significant proportion of their degree course. In order to teach in secondary schools, a large amount of the degree course must have been in National Curriculum subjects. (See profile of Jane Davies on page 85).

There is a great need for more able and enthusiastic science teachers. At secondary level there is a shortage of science teachers who are science graduates, in particular those with a physics degree. In many primary schools, science graduates have an important role supporting colleagues in the teaching of science and co-ordinating the science curriculum. Now that science is an important part of the National Curriculum for all ages, primary teachers are required to have studied science to at least GCSE grades A-C. Even so, only ten per cent of teachers are qualified in science beyond A level.

Funding for MSc and other taught courses

It is, unfortunately, a lot easier to find an interesting course than it is to obtain funding to do it. As with a PhD application, a good first degree is essential in the competitive funding situation. Some grants are available for MSc courses from the same bodies which support PhD students, although an increasing number of students on full-time MSc courses pay their own way either with the help of their families or by taking loans. Part-timers are either self-funded from earnings or supported by their employers. A PGCE (see above) is the one postgraduate course where students can get mandatory funding. To encourage more science teachers, the government is offering cash incentives; at present £5000 'golden hellos' - half to be paid at the start of the course, and the other half at the start of employment. Schools now have more flexibility in the extra allowances they can pay teachers. Because of the shortage of science specialists, science graduates can progress quickly to posts that offer these incentives.

Vocational courses

Examples of postgraduate courses leading to a professional qualification are accountancy, careers guidance, legal studies, health visiting, personnel administration and social work. Science graduates find their way into all of these fields of work. There are also postgraduate courses for people who wish to catch up in one year with what they might have done on a degree course.

Some diploma courses cover the same ground as a taught MSc but do not require training in research methods and completion of a thesis. These courses generally last nine months full-time and two years if part-time. As for other MSc courses, many people doing part-time diplomas are supported by their employer.

Courses entered by recent science graduates

The lists below give examples of the range of courses taken by postgraduates.

Graduates in maths

Biometry
Education (teacher training)
Housing studies
Hydraulic engineering
Information technology
Operational research
Social administration
Statistics
Transport engineering and planning

Graduates in physics

Accountancy
Applied environmental science
Applied optics
Computer science
Education (teacher training)
Geophysics
Fluid dynamics
Information technology
Law
Materials engineering
Mathematics
Medical physics
Operational research
Process safety
Radiation and environmental protection
Semi-conductor physics
Teaching English as a Foreign Language

Graduates in chemistry
Acoustics and noise pollution
Administration
Analytical sciences
Computing
Education (teacher training)
Information science
Law
Management studies
Marketing
Material testing
Medical chemistry
Polymer chemistry

Graduates in biological sciences
Biotechnology
Crop protection
Ecotoxicology
Education (teacher training)
Ergonomics
Export management
Forestry
Genetic counselling
Immunology
Information technology
Journalism
Marine resource development
Nutrition
Printing and publishing technology

What next after a higher degree?

The careers services in universities and colleges are not only used by undergraduates; if you are on a postgraduate course or are studying for a higher degree you'll need advice too. Most science departments have contacts with industry and with other university departments both in the UK and overseas. When you decide what you want to do they can make contacts on your behalf, particularly if you are on a specialist course. Most departments should also be able to provide information about the jobs taken up by students who have recently completed their postgraduate courses.

All full-time graduate students doing science PhDs are invited to attend graduate summer schools, usually at the end of the second year of their PhD. The courses last one week and are run by the CRAC 'Insight into Industry' programme. During the week, the participants are encouraged to consider a wide range of different careers and have an opportunity to meet recent postgraduates who are now in work.

Academic research

If you want to continue in academic research after completing a PhD, you can go on to a postdoctoral fellowship either in the UK or overseas. These fellowships are seen as a way of broadening your research experience and giving you time to find more permanent employment. British scientists are encouraged to go overseas to do this, and many do - mainly to the USA and continental Europe. Opportunities to work in a research setting overseas are predominantly in the sciences.

Funding for postdoctoral fellowships, including travel and other expenses, is available from a variety of sources: university departments, research councils, charities and European and international organisations.

Research difficulties

Career development for research scientists is always affected by the way research is funded. The Royal Society reports that there are now more collaborative projects between science research departments and industry, and that more scientists are moving into industry at later stages of their careers.

The trend in research funding in recent times has been towards short-term grants from government and other bodies, with staff working on temporary contracts. Today, in British universities, nearly half of the academic staff in science and engineering departments are on short-term contracts. As a result, the time taken for young researchers to progress to a permanent post has lengthened.

Research science can also put a great strain on young families. Because career advancement and future funding depend on the researcher's early success at postdoctoral level, long and irregular hours in the laboratory are the norm for those who want to succeed.

A few years ago, because of concern about the demotivating effects of this lack of security on research scientists, the Wellcome Trust, the Lister Institute for Preventive Medicine and the Royal Society set up schemes funding university research fellowships. Their main motivation is to provide some support and freedom for young research scientists involved in postdoctoral work.

A further aim of the schemes has been to encourage the cream of young British scientists to stay and work in Britain. This appears to

> 'The idea we try to push is that they should spend the first few years trying to establish themselves internationally as research scientists and progressively start to think of a permanent job.'
>
> *Royal Society spokesperson*

have been successful: Wellcome and the Royal Society have had numerous applicants for each place and those who have been offered fellowships have tended to stay in this country.

Although these high-flyers are a tiny minority of science graduates, their career development is vitally important to the science base in this country. The fact that some researchers take advantage of research opportunities overseas could be seen simply as an extension of the available job options, or might mean, as some suggest, that 'pure' scientific research is not valued in the UK.

For more information on opportunities overseas, see chapter 8.

Starting salaries - a postgraduate's bonus?

Advancement in some professional careers is not possible without further study, and most employers will pay a premium to graduates with a Master's or other postgraduate qualification.

A recent survey of the Association of Graduate Recruiters compared the average starting salaries for people with higher degrees with those of graduates. The average graduate starting salary for 1998 was about £16,000. Employers were paying £2000 more for a PhD (DPhil) and £900 more for an MSc or MA.

These are median figures and there is of course considerable variation, as there is among the starting salaries of graduates. Some employers pay over £3000 more, while others pay little extra. The figures refer to starting salaries in first jobs, and give no indication of future earning potential.

Making a choice

The science graduates in the profiles who did postgraduate studies had many different reasons for doing so. Jessica Harris (page 120) was encouraged to do so by her university lecturer, whereas Mike Partridge (page 118) studied part-time for his PhD while working as a research scientist with Cranfield University. Sue Wood (page 44) found employment hard to come by, so hoped that an extra qualification would give her specialist knowledge and experience, while Jane Davies (page 85) wanted to teach. Rob Brown's (page 146) career progression was undecided at the outset but doing his PhD *'has been an essential asset to getting to my current position'*.

If you are interested in postgraduate study, you should start making applications very early in the final year of your undergraduate course. Information about opportunities is available through the careers services in universities and through university departments. If you are already in employment or are applying for jobs you should also discuss the options with your (prospective) employer.

To summarise

In all, 40 per cent of graduates in physics and chemistry are likely to do a higher degree or diploma immediately after their first degree. People with first class degrees are even more likely to go on to further study.

- ☐ More science graduates than arts or humanities graduates do higher degrees either full-time or part-time.
- ☐ Computer science and engineering graduates are most likely to go straight into work.
- ☐ About five per cent of all science graduates go into teacher training. The employment prospects in teaching are good, particularly for science graduates.

- Higher degrees are taken either by teaching or research. Most diploma courses are taught courses and often have practical experience built in.
- Most employers pay a higher salary to new graduates with higher degrees, although the amount of the differential varies considerably.

Chapter 8: International opportunities

This chapter covers:
- ☐ the international nature of science
- ☐ studying abroad at different stages
- ☐ working abroad in different ways and in various countries
- ☐ the international community of scientific research
- ☐ the importance of language skills for work in the European Union.

Scientists abroad

The international nature of science means there are many opportunities for science graduates to work or study overseas at some stage of their careers. One of the stated aims of the European Commission is to enhance the mobility of scientists between member states. Multinational organisations, world trade and commerce all offer the chance to travel and work abroad.

Some young graduates choose short-term opportunities abroad while they have few responsibilities. Some want the chance to travel. Others may want to live and work abroad permanently.

At present, many young scientists in academic research are finding it difficult to get more than short-term contracts in Britain, so they are looking abroad. In the USA and, increasingly, in the rest of Europe, young British scientists are offered more job security, higher salaries and better funding for equipment. At the same time, many young scientists from other countries are coming to Britain for postgraduate and postdoctoral training before returning home: one third of all graduate students in the UK are from overseas. People also seek work abroad at a later stage in their careers because they feel that opportunities are limited in the UK.

Ways to go abroad

These can be grouped as follows:
- ☐ All or part of an undergraduate course abroad
- ☐ A postgraduate course abroad
- ☐ Postdoctoral training overseas

- ☐ A permanent job in another country
- ☐ A job that takes you overseas
- ☐ Working for a multinational company
- ☐ Government and international organisations
- ☐ Casual work and volunteer projects.

All or part of an undergraduate course abroad

Studying for a first degree at a university overseas is not a very practical idea for most people. There are numerous difficulties to overcome: education systems in other countries are very different, courses can be much longer and there will probably be no grants available - you would have to find your living expenses as well as the tuition fees. You also need to be fluent in the relevant language, although this is not such a problem in the USA, Canada or Australia!

It is much more feasible to choose a degree that allows you to spend time in a university overseas as part of your undergraduate course. There are various schemes that will enable you to do this. A few courses lead to dual and triple awards. Many universities offer science courses which include a year of study in the USA. These are popular courses and get a lot of applicants.

The Socrates-Erasmus programme has been set up by the European Commission to promote student mobility and cooperation in higher education in Europe. On the programme, students spend between one term and a year studying in a European university on a programme which has been set up by their home university and which carries academic recognition. A Socrates-Erasmus grant will cover travel and any living expenses over and above those which would have been incurred at home. Tuition fees are paid by the home university.

The Socrates-Erasmus programme also covers postgraduate study and there is a long list of affiliated university departments. Very often, these European links are for research degrees. The proportion of UK science graduates who go on to postgraduate study abroad is small for a number of reasons, including:

- ☐ the differences between the higher education systems
- ☐ funding difficulties
- ☐ lack of language skills.

If you are considering one of these programmes, it is vital to study the language of the country concerned to a standard which will allow you to understand lectures and tutorials and participate fully in student life. The Socrates-Erasmus programme can provide financial help with tuition in the language of the host country.

Finding a course

There are a number of undergraduate courses that will give you the chance to study languages as well as science. These could be combined courses in science and languages, but more often they are full science degree courses with language study alongside. Languages at A/AS level are not always required in order to do this kind of course, although clearly they would be a help. Many students complete their courses successfully with only a GCSE base in languages, but this does involve extra work.

Some students in applied science courses are lucky enough to be offered work experience overseas in their sandwich year. As part of her Biochemistry course at Bath University, Rebecca recalls:

'I spent the second of my six-month placements in a laboratory in Gottingen University doing molecular genetics - absolutely brilliant! I went there speaking no German, and I came back thinking in the language. The project was fantastic and put the whole biochemistry course into perspective. But for lack of funds, I would have returned to Germany to do a PhD.'

Taking time out of your university course for a work placement is common practice in many European countries: it is a way for employers to try out potential employees with no long-term commitment. These are called 'stage' in France and 'Praktikum' in Germany. Such placements can be obtained through university departments or through exchange programmes such as Leonardo da Vinci.

Information about Socrates-Erasmus courses can be found in the UCAS Handbook. Every year there are new schemes, and the details of the existing ones may change. You can also find out more from some of the publications in the Book List, from university departments and from university careers services. There is now much more information available through all careers services about studying and working in Europe.

British degrees are shorter and tend to cover a narrower syllabus than those on the continent. Concern about British science graduates being less competitive internationally, and at a disadvantage for postgraduate training in the rest of Europe, has led to the formation of four-year courses in physical sciences at King's College London, leading to an MSci and intended for high-flyers. In some ways they are rather like the enhanced four-year engineering courses (MEng) now on offer.

A postgraduate course abroad

Some opportunities are available through UNESCO, which produces a book on study overseas, giving details of funding arrangements. Scientists make up about a quarter of postgraduates taking UNESCO fellowships, with France and Germany being the countries taking most. Some people choose to study overseas after gaining some working experience here, for instance taking courses such as management in one of the European business schools.

Every year a number of young British science graduates go to the USA to do postgraduate training leading towards a doctorate. The structure of higher education is very different in the United States: graduate students start with a taught course before beginning their PhD research project so you must be prepared to spend longer before getting a PhD than you would do in this country. You could even find yourself alongside American students who have graduated in other subjects, including humanities, and who are starting science at postgraduate level.

The funding for postgraduate courses in the USA comes through each university department; in the private universities like Stamford and Yale, no particular preference is given to American students. There are other sources of funding such as scholarships from the Fulbright Commission. A British student with a very strong academic record has a good chance of being accepted.

Postdoctoral training overseas

The most common way for young British research scientists to get experience abroad is at postdoctoral level. Short-term posts usually last two to three years, and are an excellent opportunity to broaden your experience and establish contacts in North American and European universities and research institutes. Grants and travel scholarships are provided by the funding bodies including the Royal Society, the Wellcome Trust and NATO. For example, there are Royal Society-Fulbright Commission postdoctoral science fellowship programmes to the USA. A work permit is not required for the USA, although you must have a 'J' visa, which allows you to complete your fellowship. Competition is fierce, with applicants from all over the world.

If you want to take advantage of the international nature of research science, this is the route for you. It is an opportunity that is peculiar to science postgraduates: it is almost unknown in other academic subjects.

Scientific research

In scientific research there are opportunities in 'intra-national' collaborative research laboratories, as well as in universities and research institutions. Prestigious institutes will compete to attract the best scientists no matter what their nationality, and they will usually recruit people with a proven track record in their own country. Specialist skills can be a great advantage and scientists working in rapidly expanding areas of research, such as gene cloning or super-conductivity, may suddenly find themselves very employable in other countries.

Many academic scientists working in the UK have the opportunity to travel to conferences and make visits to other countries through their jobs.

A permanent job in another country

Is it really possible for new science graduates to get their first jobs overseas?

It happens, but is very unusual: only about three per cent get permanent jobs abroad straight after graduation. Many more new graduates take temporary jobs, and of those finding permanent jobs overseas most do so later, after they have first gained experience from a UK base. At postgraduate level, the proportion working abroad straight away is higher, not including those doing postdoctoral fellowships.

There are a number of reasons why so few graduates get their first jobs overseas.

The difficulties

☐ Higher education systems differ: degree courses are longer in the rest of Europe so most new graduates from British universities seem very young to European employers.

- The language skills of British science graduates are poor in comparison with other Europeans: a lack of understanding of different cultures can also cause problems.
- With improved higher education and training systems abroad, each country has graduates of its own and may well prefer to employ them. In non-EU countries, the employer would need to get a work permit for an employee who is not a citizen. This could prove difficult if there are suitable locals who could do the job. Unemployment exists outside the UK too!
- Many countries have tightened their regulations regarding visas and recruitment.
- A new graduate is unlikely to have much useful experience to offer. Employers in most European countries prefer to employ graduates with 'relevant' degrees. There are fewer jobs on the continent for which any degree would be acceptable, unlike the UK, where many administrative, trainee management and financial posts can be filled by any graduate.

Possible solutions

In spite of these difficulties, if you are determined and well prepared you may still succeed. There are always exceptions.

- Improve your language skills: a knowledge of other European languages is a great asset if you want to live and work abroad.
- If you have postgraduate qualifications, you are more likely to succeed.
- If you want to work overseas and do not have higher qualifications or contacts, it will pay you to get some job experience first so you have more to offer an employer.
- Links with companies through an exchange scheme, or through temporary work, a 'stage' in France or a 'Praktikum' in Germany, will give you an advantage.
- Join a British company that might offer the chance to work abroad later on, or to apply to a multinational company with branches all over the world.
- You could join the Department for International Development on their Associate Professional Officers Scheme (APOS).

Chapter 8 - International opportunities

Where are the jobs?

Some European countries have shortages of trained scientists, as well as engineers and computer scientists. However, the opening up of Eastern Europe has altered the dynamics of the labour market in Europe and scientists from the East are now competing for jobs and training in the West.

There are opportunities for experienced staff in many African countries in agriculture, engineering, science teaching and other areas. There is some direct recruitment but this is mainly through UK-based organisations such as Crown Agents or the Department for International Development.

Arab countries - particularly the major oil-producing countries who are undergoing economic expansion - advertise for staff in the European press. They usually require senior managers and qualified professional staff who have experience related to the petrochemical or construction industries. Although attractive (tax-free) salaries are offered, it is important to consider carefully the implications of living and working in a country with a very different culture.

In Australia and Canada there may be less of a cultural difference but there are fewer opportunities. Because of the difficulties involved in obtaining work permits, only those foreigners with very specialised qualifications and experience have a chance, especially during an economic recession. A British-owned company in Australia would find the same obstacles to hiring a British scientist as a US-owned company would when trying to employ an American citizen in Malaysia.

A job that takes you overseas

The UK is one of the world's foremost overseas investors and, for historic reasons, has commercial and cultural interests throughout the world. As a result, many British companies require their staff to work abroad for varying periods. Although improved communications mean fewer staff are now sent to live abroad permanently, an increasing number are going out for shorter periods from a UK base.

As UK firms begin to see themselves more and more as European rather than British, they may find it worthwhile to establish a base on the continent, with staff moving to and fro. Many British companies are now keen to provide young workers with experience abroad early in their careers.

Marketing and technical staff in particular are now more likely to be based in Britain and sent overseas on a regular basis. There are also increasing numbers of consultancies, in finance and computing for example, and more recently in the transport, water treatment and communications industries. These firms are competing for work around the world, often working for government departments. Their staff are required to travel extensively, working on short-term projects in different countries.

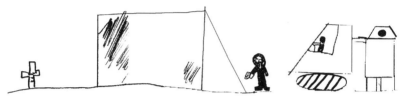

Scientist at work by Matthew

Business travel is usually highly structured and designed to fit in the maximum amount of work in the minimum time away from the home base. Your evenings may also be taken up with work-related activities and your weekends spent on planes, either moving on to the next destination or flying back home to start again on Monday morning. Travelling for work is far from a holiday: there are few opportunities to go sight-seeing on a business trip. Scientists who travel to academic conferences, however, often find that the pace is more relaxed and can be combined with a few days' leisure.

Working for a multinational company

Many young graduates apply to join a multinational company in the hope that they will have the chance to travel and work overseas. The idea of working for a multinational may sound glamorous, but staff do not always have a great deal of choice about where they work or how much time they will spend outside the UK. The opportunity to travel depends on the nature of the business and the way the firm organises staff development and promotion. Assignments vary too: a job in a Middle Eastern country is a very different proposition to a job in Europe. Language skills are always a great asset, no matter which country you visit, and can help you overcome some of the cultural differences that prevent many visitors from integrating into the local community. Language and cultural barriers often mean that social life is restricted to the company or an expatriate community within the 'host' country.

Spiralling costs have created many joint ventures between members of the European Union, and the aircraft industry is a good example of where engineers have found many new employment opportunities in Europe. Paul Pilkington (see page 89) is a physics graduate working for British Aerospace and is looking to take his career to Toulouse in France, to become a production support engineer for Airbus Aircraft.

Government and international organisations

A number of UK government departments have opportunities for people who wish to travel and work for periods overseas. The Diplomatic Service, the Ministry of Defence, the Department for International Development and the Department of Trade and Industry are obvious examples. You might also consider the British Council and other international organisations such as the World Bank, OECD, the United Nations and its related organisations.

Then there are the European Union institutions and international collaborative projects, the European Patent Office and international research laboratories. Not all these organisations specifically recruit scientists, but many science graduates have a lot to offer. Although there are not many British science graduates working for them at present, there should be greater opportunities in the future. More and more often, the focus of international diplomacy and international relations is on technical and scientific problems. Well-trained and articulate people will be needed to solve them.

Celia Keen

The chemistry graduate in this profile describes how her love of German and her keen interest in chemistry combined to give her a career which takes her abroad for part of her working time.

Career Profile
Age 35
Chartered Patent Attorney (CPA) and European Patent Attorney (EPA)
A levels: Chemistry, Physics, Maths, German
Degree: Chemistry at Oxford University (Somerville College)

'I enjoyed most subjects at O level but chose science A levels because these could lead to the greatest breadth of degree and career options. I was reluctant to give up languages though, so I took A level German too. Besides the sheer

enjoyment of this, there was the advantage that German is relevant for chemistry, which I had always found the most interesting physical science.

I read chemistry at Oxford, which has a good reputation for the subject and the added benefit of a four-year course. The fourth year is devoted to a research project which provides a gentle introduction to the world of original research. I saw this as a considerable advantage of the Oxford course since, at that stage, I did not know whether I ultimately wanted a career in research.

I enjoyed the research year and considered staying on to do a doctorate (DPhil). However, I quickly realised that research demands a narrow focus, with a high degree of specialisation in a small technical area. My own inclination, in contrast, was to seek a broader outlook. I wanted to use my chemistry but to embrace other areas of expertise too. The career of a patent agent combines science with law and also involves a linguistic component, so this seemed the ideal choice.

A patent is a national right, in respect of an invention, which can be granted under the legislation of most countries. It gives the holder the right to stop third parties carrying out the invention without permission and is generally enforceable by court proceedings. Patent agents (or patent attorneys, as we're now known) are professional legal practitioners who act on behalf of clients in all aspects of patent law. Our activities include writing patent specifications, filing and prosecuting patent applications and assisting in the defence and enforcement of granted patents. We may also advise on other aspects of intellectual property such as trademarks and copyright.

A patent attorney must have a science or engineering degree, because the ability to understand the technology underlying the invention is essential. On-the-job training in the relevant law and practice is then conducted, under the supervision of qualified practitioners - rather like an apprenticeship, in either a private partnership or an industrial company.

I joined J A Kemp & Co after finishing my degree. My training at the office was supplemented by evening lectures run by the professional Institute, and, in my second year, I studied for a Certificate in Intellectual Property Law on a full-time course run by Queen Mary and Westfield College.

I subsequently took exams to qualify as a Chartered Patent Attorney (CPA) and European Patent Attorney (EPA). I am, therefore, entitled to practise before the UK Patent Office and the European Patent Office, which is based in Germany. The latter is a trilingual organisation and EPAs are required to have a knowledge of English, French and German. The qualifying exams, therefore, include a language component, so my German was very useful!

The work which I handle is all chemical or pharmaceutical, and very varied. I deal each day with inventions relating, amongst other things, to drugs, catalysts, polymers, printing inks and chemical processes. My clients are mostly British and foreign companies, though, so I don't tend to encounter the archetypal 'garden shed' inventor!'

Casual work and volunteer projects

Before they decide on a particular career, many students are keen to spend some time after they finish their degree courses living and working abroad. If you would like to see more of the world in this way, there is a wide range of routes you can take.

Lots of new graduates take short-term casual work, and scientists are no exception. You could experience life on a kibbutz in Israel, go to Australia on a 'young person's' visa, work in summer camps in the United States, help on an international work camp or take part in a project run by a group like Christians Abroad.

Voluntary organisations are usually looking for people with special skills and often ask for a longer term commitment. In developing countries there are often opportunities for agriculture graduates, medical staff and engineers. New maths and science graduates are also sought for teaching jobs.

Voluntary Service Overseas (VSO) is probably the best known of these groups. VSO aims to promote and support human development through its work, to improve people's education and health, their income and employment opportunities, and their ability to contribute to the society in which they live. Volunteers work alongside local people - teaching, sharing ideas and exploring different ways of working and training.

As a volunteer you could end up working in a remote area so you need to be pretty self-reliant, although you will be thoroughly briefed in the UK before you go. Relevant work experience is in great demand, as most projects aim to provide local people with self-sufficiency and independence through practical training.

An overview

At every level, the international nature of science offers graduates a wide range of options.

If you decide to go into a scientific job you will be aware of scientific developments in other countries, from your first degree right through your career. You will communicate with other science specialists all over the world, through computer networks, the world wide web, conferences, visits, study and research projects overseas.

In scientific research there is an international community in each area of specialisation, and there are always opportunities for high flyers in laboratories and research institutes around the world.

Science graduates go into many other jobs with an international dimension, involving travel or working for periods overseas. For most people, these opportunities come after first gaining some experience in the UK.

This may sound as if a science degree alone is a free ticket to travel the world, but there are other factors involved. Jobs abroad often require additional skills and experience and there will be keen competition from scientists from other countries. Having said that, there are many ways in which you can enhance your chances of working overseas.

British scientists have a distinct advantage in that English is the international language of science. However, although your scientific peers may understand written English and be able to conduct technical discussions, companies on the continent operate mostly in the language of the country where they are based. So for many science-related jobs and other careers that take you overseas, language skills are a great asset.

Many science graduates work overseas for some of their working lives, and some emigrate permanently, living and bringing up their families in a different culture. If an idea of an international career appeals to you, there are many ways in which you can prepare to become a world citizen. You'll have a high level of expertise in a field that is in demand, together with a working knowledge of at least one other major language, French or German for example, and the desire to experience a new way of life.

Chapter 9: Your future as a science graduate

This chapter covers:
- ☐ where science graduates go after their first jobs
- ☐ careers in research
- ☐ changes in career development for science graduates
- ☐ planning your moves
- ☐ future demand for science graduates.

Vital statistics

As we have seen, the first destination statistics give only a limited view of the careers followed by science graduates. This book has shown you a selection of the wide range of first jobs, but these are just the starting points. People change jobs and develop in many different ways. For instance, more and more young graduates are taking temporary jobs or going on to further study. In this chapter we will be looking at what happens afterwards.

Information on the career progression of science graduates is much harder to obtain because people move around and the careers service loses touch with them. However, a 1995 survey of graduates of Sussex University (1991-1993) by the Institute for Employment Studies and the University's Career Development Unit found that:

Scientist at work by Louise

- ☐ applied science (which includes engineering) graduates were almost twice as likely than others to have been in continuous permanent employment over the first three years

☐ graduates from applied sciences had less changeability.

And in the follow-up study in 1997:

☐ science graduates, especially those in biological sciences, were more likely to be in further study; engineering graduates were by far the more likely to be in jobs

☐ the highest salary earners were graduates in engineering and mathematical science.

In the IES Annual Graduate Review 1998-1999, graduates in computer science, engineering and maths have been the most likely to move into high level managerial, professional or technical jobs, contrasting with those with degrees in biological science which have amongst the highest initial unemployment rates, and who are the least likely to be in the higher level jobs. Over 400,000 students graduated in 1998 - more than double the numbers of 10 years ago, although entry to physical sciences, engineering and technology has been falling in recent years.

However, the employment situation changes very quickly, and surveys and statistics which you can refer to now may be completely out of date by the time you get to that stage. Alterations in corporate structures, economic and technological developments, political changes in Britain, continental Europe and the rest of the world are all factors that affect the supply of jobs and the lives of the people doing them. The people who are in senior positions today probably started out 30 or 40 years ago. Things were very different then, and they will have changed again by the time you get there. Above all, be flexible! As Garry has found out during his varied and challenging career so far (profile page 78) '*today's employment world is constantly changing and you must be flexible and acceptant of change if you are to survive*'.

Scientists in Britain

Do science graduates get the top managerial jobs in Britain the way they do in other European countries like Germany and France? The answer appears to be that too few hold senior positions. The reason for this may lie in our ambivalent attitude towards science and scientists. Winston Churchill has often been quoted as saying, '*Scientists should be on tap but not on top!*'

It is possible to find people at the very top of British companies with science and engineering qualifications, but there are many more with financial or legal training. In Germany, by contrast, many bosses in industry are professional scientists. For instance, the position of Federal Minister for Research and Technology has also been held by a scientist - Dr Heinz Reisenhber. Over here, despite the fact that we had a prime minister who bucked the trends by being both a woman and a chemistry graduate, there are still very few scientists in the higher ranks of the Civil Service or in parliament. It seems that Margaret Thatcher was quite unique.

The position of graduates

The Association of Graduate Recruiters, an organisation that represents the interests of employers who recruit graduates, asked its members at what level most graduates work in their companies. Surprisingly, the response indicated that graduates predominate in most of the professional and managerial levels except at the very top; in contrast to the position in Germany, France, Japan and the USA. If graduates in Britain are not yet dominant in top management, science graduates will not be there either. However, graduates in Britain are now better represented further down the corporate hierarchy, so, in time, the balance should alter higher up the ladder.

Unfortunately, more graduates in senior management does not necessarily mean more science graduates at the top. It has been a concern that science graduates could get stuck in research and development jobs, with few opportunities to progress to general management. A young science graduate often joins a large company straight into a specific job, whereas others join a general management training programme that often involves working in different areas of the company. Science graduates could find themselves cut off from the rest of the company, while others are more aware of the whole organisation, which gives them an advantage when it comes to promotion up the management ladder.

However, science graduates starting in science research work can progress into other science-related activities as their careers develop. The choices vary, depending on the industry and the particular company you are working for.

Rob Brown

Rob's career has taken him from a government-related establishment into a commercial organisation.

Career Profile

Age 31

Senior Molecular Biologist

Biocatalysis Group, Chirotech Technology Ltd, Chiroscience Group plc, Cambridge Science Park

A levels: biology, computer science, design and technology

Degree: BSc Science and the Environment (Biotechnology) Leicester Polytechnic

PhD: Molecular Biology and Pathogenesis, Open University. Based at Centre for Applied Microbiology, Porton Down.

'After drifting through sixth form with little career direction (my A level subjects being testament to this!) I decided, at the last moment, between a managerial training job with the AA or to go through clearing and find a course at a polytechnic or university. I couldn't get into university so the Friday before term started, I managed to wangle a place at Leicester Poly. The course comprised many facets: from computing and maths, through environmental economics and ecology to toxicological regulation and biotechnology. Presented with all these disciplines, I thought I was still keeping my options open for my future career, wherever it lay.

I found the first year particularly difficult, struggling through chemistry, having to re-sit the exam at the end of the summer. The second year was not so bad; the second semester specialised in the biotechnological side and not the more mundane ecology.

At the end of my second year, I opted to take a 12-month industrial placement at the Centre for Applied Microbiology and Research (CAMR). This establishment was part of the Public Health Laboratory Service (PHLS), specialising in containment and research into bacterial and viral pathogens. During this time, I gained experience in a wide range of molecular biological and immunological techniques, and managed to publish some of my work in a scientific journal.

I found this experience invaluable upon my return to Leicester to complete my final year. Indeed, the experience and links I made at CAMR were sufficient to secure me a job and the opportunity to study towards a PhD at the same establishment where I graduated.

On returning to CAMR, which was now a self-funded authority under the auspices of the Department of Health, I was employed as a research scientist within the Biologics Division, responsible for cloning and expressing bacterial toxins for vaccine and therapeutic development. Some of this work also went towards my PhD study, which was registered through the Open University. After three to four years' research, I submitted my thesis and continued at CAMR within the same field.

After a total of five years, I realised that it was time to move on. I wanted to expand my experience outside a government-related establishment into a true commercial environment. I applied for a job at Chiroscience Ltd, a small but growing pharmaceutical company on the Science Park in Cambridge. Indeed, Cambridge has the highest concentration of biotech companies in Europe, and, with its affiliation to Cambridge University and other world-renowned academic institutions, I thought this was a good place for a relatively fresh post-doctoral scientist to begin his commercial career.

The field of biotechnology is innovative and very dynamic. Before I had been with the company for six months, things were changing. Chiroscience has expanded its interests into three separate companies within the Chiroscience Group plc. Presently, I work for Chirotech Technology Ltd., a speciality chemicals company providing intermediate compounds to the pharmaceutical industry.

Since leaving sixth form 12 years ago with no career direction, I have completed two degrees. Although my career does not appear to have any purposeful development, doing my PhD has been an essential asset to getting to my current position. I thoroughly enjoy my work and hope to take advantage of many of the opportunities that arise in this precarious, but exciting, industry.'

Progression from research into other science-related activities might include the following options:

Research management
Development work
Project management
Production management

Quality assurance
Marketing - market assessment
Licensing of intellectual property
Commercial development
Information technology.

All of these activities could lead you into middle and then senior management. Once again the move will be from specialist jobs to generalist positions, rather than the other way round. Science graduates are unlikely to start in marketing and move into research and development, or to start in personnel and move into production. People with scientific and technical specialisms have to learn new skills in areas like management, finance, commerce and marketing.

Career progression and hierarchy

The corporate structure itself is also altering. The Association of Graduate Recruiters' report *Recruitment of Graduates in the 21st Century* summarises the changes.

- Organisations are now 'flatter' and less hierarchical, with layers of middle management removed.
- There is a growing need for specialists educated to a high level, who will tend not to be managers of other staff.
- Advancement no longer means simply climbing up the organisational ladder.
- Graduates are expected to assume more responsible roles sooner than they would have done a decade ago.
- Service industries and the professions are re-thinking their productivity and what they will require for both recruitment and retention of staff.

These changes are welcomed by many science graduates. More young families have two graduate parents, both of whom wish to pursue a career. The opportunity to continue in a specialisation without feeling the pressure to go for career advancement and join a management hierarchy is seen as a great advantage. Layers of expensive management can lead to frustration for the scientists on whose creativity companies depend. The move towards less formal working relationships fosters that scientific creativity and helps people to balance their professional roles more comfortably with family life.

Planning your moves

The difficult thing for young scientists is to decide when the time is right to move out of research, if this is what you eventually wish to do. There is a danger of leaving it too late. It is very difficult as a new graduate to know which way your skills and interests might lead you at different stages of your career. Research in industry attracts a very different individual from the one who prefers development work - turning ideas into potential products, overseeing pilot production testing and scaling up. Some science graduates do get more interested in development work later on in their careers, but others decide to move into quite different areas.

A crucial element of career planning is avoiding the cul-de-sac, unless, of course, it is one you are particularly happy in. Being able to move to other areas, using your expertise in a different setting, learning new skills, having the chance to try new roles - these all seem more desirable in a career than edging up predetermined structures and waiting around for 'dead men's shoes'.

And you have plenty of expertise already. There are many skills that you have acquired through academic study, and through working in research and development, which are easily transferable to other career areas, and much in demand.

- Numeracy - scientific disciplines involve working with numerical data at a level which is acceptable for most jobs
- Communication
- Thoroughness and accuracy
- Problem-solving
- Self-motivation
- Organisation and time management
- Team skills
- Information technology skills

Responsibility for your career planning is, ultimately, your own, although many companies encourage their staff to think of career progression outside research and development as well as within it.

Industry is not the only option

Not all science graduates want to run large industries or stalk the corridors of power. Many want to work in some aspect of science because it interests them and they feel it is worthwhile. They feel they can make important contributions to the development of knowledge, to education, to the application of science to problems for the benefit of our own standard of living and economy, as well as to global problems such as poverty, hunger and the environment.

Some science graduates see themselves as scientists all their working lives. Others use their scientific training as managers, teachers, librarians, charity field workers, publishers, agricultural advisers, actuaries or patent agents. Different roles suit different people at different stages of their careers: the job you choose early on in your career may not be the right one in a few years' time. By staying flexible and maintaining your transferable skills you can give yourself more options when the time comes to reassess your situation at work.

There continues to be a severe shortage of science teachers in secondary schools, particularly in physics, and a lack of scientists entering teacher training courses. Institutions running PGCE courses can offer means-tested extra funding through the Secondary Subject Shortage Scheme. At present the Government is offering a £5000 financial incentive, paid in two instalments, to encourage students to train as maths or science teachers.

Where will the jobs be?

In academic science, shortages are predicted. If young academics are not recruited now, there will be unfilled vacancies in higher education when the present generation retires. There will be a continuing demand from the schools and from further education as the importance of science and technical education is more fully realised.

The general public's increasing interest in science and science-related issues means more opportunities for science graduates in the media and in communications. The continual growth of scientific knowledge requires explanation and interpretation: this will lead to jobs in areas like specialist communications, information science and technical publishing.

Information technology will continue to dominate communications at all levels of industry, commerce and public and private life. Again scientists are in a strong position to find opportunities here. The IT knowledge and experience gained on your degree course will be a great asset in all sorts of jobs and will give you a real advantage over less IT-literate graduates at all levels. In 1997, the biggest shortages of skilled graduates were in information technology, science and research and development.

An area where scientists are likely to be in demand is in patent work. (See profile of Celia Keen on page 139). Although the route to qualification is rigorous, and often requires study at postgraduate level, patent agents are rarely unemployed, and it is not unheard of for graduates with only two or three years' experience to be attracting salaries of up to £40,000.

The internationalisation of science-based companies will continue to increase opportunities in the EU and further afield. Opportunities will continue to be offered in developing countries where growth depends on scientific expertise. The peculiarly British phenomenon of having few scientists in top government and management posts may be changed by influence from countries that are more successful in science-related industries.

In the near future, science graduates may find themselves in increasing demand because the proportion of people studying science is not keeping pace with the general rise in the number of graduates. Competition for jobs that require a science degree will not be as high, and the articulate and literate science graduate may have a 'scarcity value' when applying for the whole range of other jobs open to graduates of any discipline.

In summary

☐ Continuing changes in graduate employment mean that what happened to those who left higher education in the past is not necessarily relevant to you today.

☐ Opportunities for science graduates are increasing due to rapid developments in many areas: science, technology, communications, company structures, international relations and global markets.

☐ Science graduates (including doctors) have a wider choice of jobs overseas than other graduates.

Careers with a Science Degree

- Science graduates have an important part to play in many areas of work. To take advantage of all the options on offer, you need to complement your qualifications in science and IT with a range of skills in communication.

- An increasing recognition of the importance of science to the prosperity of this country should mean better opportunities for science graduates in the future.

Chapter 10: Where do you go from here?

This chapter covers:
- ☐ taking stock of where you are
- ☐ planning your next step
- ☐ sources of information and advice
- ☐ events you can look out for
- ☐ action you can take.

Your career, your decision

This final chapter is about what you can do now to help yourself take decisions, make progress and turn your plans into reality. You will find lots of suggestions of things to do, people to talk to and information to use. You need to think about your own interests and preferences as well as look at the different options that are available.

If all this sounds like a lot of hard work, remember it is vitally important to you. You will find so much of your time gets taken up with all the short-term issues: this week's homework, next week's concert or party, the next match, holidays or exams at the end of the year. Your degree and career choices are much more important in the long term and really deserve your attention. It is your future and only you can make it work.

Start from where you are now

The sections below will give you useful leads. If you have already started your A levels, Highers or BTEC/GNVQ course, skip Stage 1 and go straight to Stage 2.

Stage 1 - Choosing A levels, Advanced Highers, BTECs or GNVQs

Start thinking and planning at the end of year 10 and the very beginning of year 11. Use some of your holiday period too! Look at the columns below. Plan how you could achieve these aims.

Consider all your options
- ☐ Use the careers library

- ☐ Use occupational information databases
- ☐ Read *Which A levels?*
- ☐ Look at GNVQ courses
- ☐ Get prospectuses from local colleges and sixth forms
- ☐ Visit open days
- ☐ Talk to your teachers
- ☐ Talk to sixth-form and college staff
- ☐ Look at *What Do Graduates Do?*

Decide what really matters to you

- ☐ Talk to your family and friends
- ☐ Talk to your careers adviser
- ☐ Talk to your teachers
- ☐ Use computer guidance programmes
- ☐ Try to get work experience or do work shadowing
- ☐ Attend lectures and visits, science fairs and science conventions
- ☐ Read *New Scientist* and science articles in the press
- ☐ Watch programmes like Tomorrow's World and Horizon

General National Vocational Qualifications

Don't forget there are alternative qualifications to A levels which can lead on to degree courses. GNVQs have become a popular alternative to A levels and are widely accepted for entry onto higher education courses, although additional units or an accompanying A level may be required. GNVQs have a very different approach from A level with much more project work. N.B. In Scotland, GSVQs are part of a system of National Qualifications and are gradually being phased out. The National Qualifications are currently being revised.

BTEC National Diplomas

BTEC National Diplomas can also lead on to degrees courses. They are the equivalent of two A levels/Advanced GNVQ and are widely accepted by HE institutions. Distinctions/merits are generally required. BTEC NDs are mostly college-based, and last for two-years, full-time.

Maths (again!)

Take stock of your maths. For further study in science, maths is a vitally important tool. Even if you are not intending to study maths for A level, you should try to take it as far as you can. You could take AS level maths alongside A levels or BTEC/GNVQ.

Using careers software

There are literally dozens of careers software products on the market and, with the increasing availability of information on the Internet, there are arguably more 'attractive' and more popular ways for you to access careers information than ever before. There is no one program on the market which will satisfy all users from age 15 upwards, so choose one which is most appropriate for the stage you are at in your education.

Use occupational information databases to get quick and easy access to:

- work details
- entry routes and training given
- personal qualities and skills checklists
- similar job areas
- professional bodies to write to
- higher education information
- addresses for further information
- references for further exploration.

Examples of this sort of database include:

- KeyCLIPS from *Lifetime Careers Publishing*
- CID from *Careersoft*
- Explorer '99 from *JIIG-CAL Progressions Ltd*
- Kudos and Careerscape from *CASCAID*
- Odyssey from *Progressions Ltd*

Most of these have easy-to-use searching routes, enabling you to match your subject interests, perceived skills and personal interests to career ideas.

Use careers guidance and self-assessment software to help you become aware of your own occupational interests. Most will ask you to complete an interest questionnaire or guide. Your responses will be matched to a database of occupations with most programs giving you an option to see the pros and cons (likes and dislikes) on any job on the list. Some produce job ideas based on the Connolly Occupational Interests questionnaire.

Examples of this sort of guidance software include:
- Pathfinder from *JIIG-CAL Progressions Ltd*
- Kudos and Adult Directions from *CASCAID*
- Iscom and Iscope from *ISCO*.

Use databases of further and higher education to search for and identify relevant courses of study. Information can be accessed by full-time/part-time/correspondence/geographical location/A level points and so on. Gateways to the Internet are also possible on some programs.

Examples of this sort of software include:
- Studylink from *Learning Information Systems*
- UK Course Discover from *ECCTIS 2000*
- Higher Ideas from *Careersoft*
- Discourse from *ISCO*.

A number of software programs will help you find out about work and study opportunities in **Europe.** You should not underestimate the potential growth of jobs in scientific and technical occupations that are available in Europe and worldwide.

Some programs which could help you include:
- Europe in the round from *Vocational Technologies Ltd*
- EXODUS from *Careers Europe*
- ICDL distance learning database from *ICDL*.

The media

New Scientist is a weekly science magazine available in newsagents, local libraries and many schools. It has lots of news about developments in sciences and some careers articles as well. *Focus* is a science magazine that is published monthly. *The Scientific American* has much longer in-depth articles about scientific topics but can be understood by A level students. Other useful magazines include *Nature* and *Science*. Reading about science and watching TV science programmes will also help you to identify where your scientific interests lie.

Stage 2 - choosing a science degree

You should be planning and thinking soon after you start your post-16 study. So you need to start straight away. Set some of your summer holiday aside for this!

- ☐ Consider all your options
- ☐ Use your careers library
- ☐ Use computer databases
- ☐ Look at *Which Degree? Sciences*
- ☐ Look at Hobsons *Degree Course Guides*
- ☐ Look at the AGCAS Signpost series of booklets, e.g. *Your degree in Life Sciences* and others
- ☐ Visit higher education conventions and fairs
- ☐ Go to open days
- ☐ Go on introductory courses in universities, e.g. WISE (Women into Science and Engineering) and Insight into Engineering
- ☐ Look at *What Do Graduates Do?*

Decide what really matters to you

- ☐ Talk to friends and family
- ☐ Talk to your careers adviser
- ☐ Talk to your teachers
- ☐ Try to do some work experience or work shadowing
- ☐ Use computer-based guidance systems
- ☐ Use computer based HE choice systems
- ☐ Talk to scientists you meet
- ☐ Keep reading magazines like *Nature, New Scientist, Focus* and *The Scientific American* and watching TV programmes such as Horizon and Tomorrow's World

Alternative courses

Don't forget you could do a sandwich course or get work experience as part of your degree course. You might also consider a course with a European link or one which will give you the chance to study or work overseas.

Alternative qualifications

Remember there is a whole range of Higher National Diploma courses with entry requirements of one A level pass or equivalent. This could be a good alternative to a degree course if you feel the required grades are too high for you, or you prefer the more applied vocational nature of an HND course.

Work experience and work shadowing

Many schools have work experience programmes in year 10 or year 11 which give you the chance to work in an organisation and find out about the jobs people do. If you are in the sixth form you may also be able to do work experience, but on a very part-time basis or in the holidays.

Scientist at work by Peter

Work shadowing is when you spend a day or two following someone doing a particular job, watching them at work, and seeing and hearing about the job at first-hand. This is often a good way to get to know about the tasks involved in higher-level jobs which you would be less likely to be able to try out on work experience.

Ask your school or college careers adviser or your science teachers about these schemes. Girls who would like to meet, or shadow, women science or engineering graduates can make contact through the local branch of WISE. Ask your science teachers to help you.

Higher education conventions

Many careers services and UCAS run events where you can meet representatives from universities and colleges from all over the country who will tell you about their courses. You can also get information about career opportunities after courses. These are well worth attending, particularly if you have done some preparatory work first, and have prepared questions to ask and interests to follow up. You will also be able to get information about loans/finance and sponsorships.

Open days

Lists of open days in universities and colleges should be available in your school. Some are for all students, others are for students interested in particular subjects, e.g. sciences or engineering. Ask your careers teacher for information.

'Taster' courses

Some universities and other organisations, such as the Institute of Physics, run courses of up to a week for students interested in these careers. Again your school or college will have information.

Reference books

There are many reference books about higher education and the courses on offer. These will be available in your school or college careers library, in the careers centre or local public library. See the Book List for more details.

Stage 3 - moving on from your degree or diploma course

This stage may be a long way in the future, but the process is the same as for Stages 1 and 2. The skills you learn now in researching options and using careers information will help you through Stage 3 too.

You would start thinking and planning Stage 3 in your second year on a three-year course, or the third year if you are on a four-year course.

- ☐ Consider all your options
- ☐ Use your university or college careers service
- ☐ Look at *What Do Graduates Do?*
- ☐ Look at the AGCAS publication *After Your Degree/HND ... What Next?* for your subject area
- ☐ Talk to your tutors and lecturers
- ☐ Attend careers events and employers' presentations
- ☐ Look at *Graduate Studies*
- ☐ Talk to employers while on work experience or holiday jobs
- ☐ Keep reading the science magazines *New Scientist* and *Scientific American*

Decide what really matters to you

- [ ] Talk to university or college careers advisers
- [ ] Use computer-based guidance systems
- [ ] Talk to your lecturers about your work
- [ ] Talk to postgraduate students
- [ ] Seek out people doing jobs you are interested in
- [ ] Use the world wide web to investigate companies and vacancies

For women

Seek out your local WISE group and attend their meetings

Postgraduate courses

Don't forget there are all sorts of different ways of getting postgraduate qualifications. They can be taken immediately after your first degree, but also later while you are at work or after some work experience. There are part-time as well as full-time courses.

Careers services

Universities and colleges of higher education have careers services where you can have access to an information library, computer databases, vacancy information, computer-aided guidance and personal guidance from a careers adviser.

Careers fairs

Your careers service will organise presentations by employers about their companies, as well as general talks on careers topics. There may also be careers fairs, with many employers looking to recruit suitable graduates. This is where you can go and meet employers informally.

Holiday jobs

Science graduates are particularly successful at getting vacation work, usually in laboratories. This experience will give you a financial boost but will also provide some insights into jobs and how you fit in at work.

Careers courses

You may have the chance to do a CRAC Insight into Management course, which will give you the chance to work alongside young managers from industry, commerce and the public sector. It is a good way to find out about how people set about different jobs.

Further study

If you are interested in further academic study then talk to your tutors and lecturers. They will have many contacts with other departments and research groups. You will need to get information about funding too.

Investigating other issues raised in this book

The sections above have covered ways of finding out about the range of options open to you at each stage and how you can get help with your decision making. But you might also want to investigate further some of the important issues we have raised about science and science careers before making your choices.

You might want to follow up the issues of image, status, gender and career development. There is some factual information available on these topics, but you will also come across a great deal of opinion and strongly held views. Take care to apply the same rigorous scrutiny that a scientist would apply to a scientific question!

In your research you will need to examine lots of different sources and talk to many different people. Try to relate the opinions you hear to the experiences of the people you are talking to. What is their standpoint? You will need other points of view to get a balanced picture. Allow for human error and then talk to a few more people.

You will also want to think about how the information will affect you. Will you have the same experience? Do you have the same concerns? How will you avoid the pitfalls and make the best of the opportunities? Bear in mind that things are constantly changing.

You will meet useful people to talk to through everyday student life, by using the careers guidance services, through holiday jobs and work experience, and through your friends and family. Get in touch with professional associations, such as the Institute of Physics, and local firms. It is surprisingly easy to find the contacts, what takes more initiative is to ask people for some of their time and make good use of it.

You will find that most people will be happy to talk to you about their work and how they got into it. Sometimes the only problem is stopping them talking! If you are well prepared with questions on issues that are important for you, they will be glad to give you the time.

A word of warning: beware of people who offer you directive advice! They are usually thinking of themselves, not you. Advice that begins, 'If I were you ...' usually means 'I wish I had done this.' They may be trying to be helpful but the right decision for them would not necessarily suit you.

Whatever stage you are at make yourself a plan of action. Make a list of things you have to do to get all the information, advice and experience you need to give you a solid base for your next decision point. Make a list and a timetable of actions. Then make sure you follow it up. It helps if you give yourself target dates and get a friend to check your progress. It's your future and only you can make the decisions, so put yourself firmly in charge.

And finally

This book has tried to give a positive but realistic picture of career opportunities with a science degree. An education in science is not a panacea for all. The bad news is that science has, in some respects, a poor image. Scientists often feel they have low status and that they get unfairly blamed for some of the problems of today. Science graduates are not yet in enough of the very top jobs, research scientists are struggling for funding and a clear career structure, and many women scientists feel that science is still a man's world.

The good news, on the other hand, is that a science education will help you to understand and contribute to many of the most important issues of the 21st century. You will have knowledge and experience that can be applied in many different work situations, scientific and non-scientific, in the UK and overseas. You will be well equipped to deal with the information technology revolution, and you will have an international outlook appropriate for the important challenges of the future. Together with other young science graduates, male and female, in all sorts of different roles, you can make a difference. Go for it!

The final comment comes from a teacher of one of the most famous scientists of all time. It's worth remembering when you hit a difficult patch in your studies or career.

'He will never amount to anything.' Albert Einstein's school report.

Book list

Choosing A level, GNVQ and equivalent courses

Decisions at 15/16+ published by Hobsons

Which A levels? and *GNVQ: is it for you?* and *15+ Pathways to Success* published by Lifetime Careers Publishing

Great Careers for People Interested in Science and Technology published by Kogan Page

Choosing a science degree

Which Degree - Sciences, Medicine, Mathematics published by Hobsons

Which Degree - Engineering, Technology, Geography published by Hobsons

Degree Course Guides published by Hobsons

University and College Entrance: The Official Guide published by the Universities and Colleges Admissions Service

High Flyers published by Hobsons

Socrates-Erasmus: the UK Guide published by the Independent Schools Careers Organisation

Postgraduate courses

POSTGRAD: the Directory of Graduate Studies published by Hobsons

Postgraduate Study Abroad, published by UNESCO

Careers with a science degree

What Do Graduates Do?, published by AgCAS

Careers Information Booklets, published by the Association of Graduate Careers Advisory Services

Hobsons Casebook Series - Science

For women considering a career in science

Cracking it! Helping Women to Succeed in Science, Engineering and Technology by Josephine Warrior, available from Training Publications Ltd, PO Box 75, Stockport SK4 1PH.

Publishers addresses

Hobsons Publishing - titles are available from Biblios Publishers' Distribution Services Ltd, Star Road, Partridge Green, West Sussex RH13 8LD. The telephone order number is 01403 710851

AgCAS - CSU Ltd, Despatch Department, Prospects House, Booth Street East, Manchester M13 9EP

Independent Schools Careers Organisation - ISCO Publications, 12A Princess Way, Camberley, Surrey GU15 3SP. Tel: 01276 21188

Kogan Page - 120 Pentonville Road, London N1 9JN. Tel: 020 7278 0433

General science interest (All paperbacks)

The Cartoon Guide to Physics - published by HarperCollins, 1992

James Gleick, *Chaos* - published by Abacus, 1993

James Watson, *The Double Helix* - published by Penguin, 1970

Stephen Jay Gould, *Eight Little Piggies* - published by Penguin, 1994

Paul Davies, *The Mind of God* - published by Penguin, 1993

Richard Dawkins, *The Selfish Gene* - published by OUP, 1989

Susan Blackmore and Richard Dawkins, *The Meme Machine* - published by OUP, 1999

Dava Sobel, *Longitude* - published by Penguin, 1996

Lederman & Teresi, *The God Particle: If the Universe is the Answer, What is the Question* - published by Delta, 1994

Stephen Hawking's Universe: The Cosmos Explained - published by Basic Books, 1998

Useful addresses

Royal Society of Chemistry - Burlington House, Piccadilly, London W1V 0BN. Tel: 020 7437 8656.

Institute of Biology - 20-22 Queensberry Place, London SW7 2DZ. Tel: 020 7581 8333.

Institute of Physics - 76 Portland Place, London W1N 3DH. Tel: 020 7470 4800.

The Institute of Mathematics and its Applications - 16 Nelson Street, Southend-on-Sea, Essex SS1 1EF. Tel: 01702 354020.

EMTA - Vector House, 41 Clarendon Road, Watford, Hertfordshire WD1 1HS - for information about science taster courses.

Royal Society website: www.royalsoc.ac.uk

- for information about the Committee On the Public Understanding of Science (COPUS); Royal Society lectures; 'New Frontiers in Science' exhibition etc

Supporting Women

Women Into Science and Engineering - The Engineering Council, 10 Maltravers Street, London WC2R 3ER. Tel: 020 7557 6436.

Promoting SET for Women - UG.88, 1 Victoria Street, London SW1H 0ET. Helpline 020 7233 0743.

Unit for Policy Research in Science and Medicine - Wellcome Trust, 210 Euston Road, London NW1 2BE. Tel: 020 7611 8389.

The Athena Project Office - 15 Prince's Gardens, Exhibition Road, London SW7 2QA. Tel: 020 7594 5509.

Postgraduate research rating

HEFC website: www.niss.ac.uk/education/hefc

Possible sources of funding for postgraduate research

Biotechnology and Biological Sciences Research Council Tel: 01793 413200. Website: www.bbsrc.ac.uk/

Engineering and Physical Sciences Research Council Tel: 01793 444000. Website: www.epsrc.ac.uk/

Natural Environment Research Council Tel: 01793 411500. Website: www.nerc.ac.uk/

Particle Physics and Astronomy Research Council Tel: 01793 442000. Website: www.pparc.ac.uk/

All the above are based at Polaris House, North Star Avenue, Swindon.

Medical Research Council - 20 Park Crescent, London W1N 4AL. Tel: 020 7636 5422. Website: www.mrc.ac.uk/

Teaching Company Directorate - Hillside House, 79 London Street, Faringdon, Oxfordshire SN7 8AA. Tel: 01367 245200. Website: www.tcd.co.uk/ (for research assistantships)

Career Development Loans - call 0800 58 55 05 and quote HSG/C/1

European funding - consult the website: www.europa.eu.int/ and use the CORDIS link.

Glossary of science courses

This glossary gives a brief description of the main science degree subjects offered at universities and colleges. There are many other degree courses with variations on these titles. Sometimes this is because the course deals with a specialised branch or application of the subject and sometimes because the course approaches the subject in a particular way. For example, many degree courses have 'applied' in their title. These courses usually cover much the same basic theory as the 'pure' courses, but place more emphasis on the applications of the subject and may also include periods of industrial or laboratory experience outside the university.

The title of a course is only a rough guide to what it actually contains, as there can be wide variation in the content of courses with the same title and considerable overlap between courses with different titles. Fuller information about individual courses is available in the Degree Course Guides, Which Degree series (both Hobsons), and from university and college prospectuses.

All university and college degree courses in science include some mathematics, statistics and computer methods. Most courses also have practical laboratory work and, where appropriate, fieldwork; nearly all courses allow for extended project work in the later stages. Practical work is used to reinforce theoretical work as well as to teach practical techniques and experimental design.

Acoustics - the science of the production and transmission of sound, and its behaviour when it is reflected or absorbed by surfaces. Acoustic engineers are involved with both the enhancement of wanted sound and the suppression of noise that is unwanted or environmentally damaging. There are important applications in architecture and building, and in areas of mechanical engineering such as aeronautical and automotive engineering, as well as in the music and entertainment industries.

Agricultural Science/Agriculture - courses normally cover a mixture of the scientific, technological, environmental, practical and business aspects of agriculture, though there is considerable variation in the emphasis given to each component. Several courses require students to have some practical experience of agriculture before the course begins and all have a strong practical element through laboratory and fieldwork projects. The science content is based on chemistry, biochemistry and plant and animal biology. As the courses progress, topics with a more specific application to agriculture are introduced such as genetics and plant and animal breeding, animal nutrition and physiology, parasitology, stock husbandry and crop and soil science. Agriculture rapidly finds applications for new discoveries, so topics and specialised degrees in areas such as agricultural biotechnology are beginning to appear. Specialised degree courses are available allowing you to study a particular branch of agriculture such as crop or animal science or production, forestry or horticulture, or the business aspects. Other courses concentrate on a specific area of agricultural science, such as agricultural microbiology, usually with less emphasis on the practical farming content.

Agroforestry - the combined study of agriculture and forestry. As well as studying components of agriculture and forestry separately, there is emphasis on how mixed farming and forestry systems can be sustained in what are often environmentally sensitive areas.

Anatomy/Anatomical Sciences - the study of the structure of living organisms (although the term plant anatomy is widely used, degree courses in anatomy and anatomical sciences are concerned almost exclusively with mammalian anatomy and in some cases only with human anatomy). Courses have changed over recent years with much less time devoted to the description of body structure (topographical anatomy). There is now much greater emphasis on the relationship between structure and function and on the way molecular and cellular structures determine macroscopic properties and structures, such as the organisation and function of the body's organs, skeleton, nervous and other systems. Specialised topics within anatomy include histology, embryology (the study of development from the fertilised egg), pathology and the comparative anatomy of different species.

Anatomy has links with biochemistry, physiology, genetics and microbiology. Anatomy is also an important component of courses in medicine, dentistry, veterinary sciences, human biology and most of the professions allied to medicine such as physiotherapy and radiography.

Animal Science - see zoology

Arboriculture - the breeding and cultivation of trees and shrubs including planting, pruning, felling, prevention and treatment of diseases and protection from pests. Degree courses in forestry include arboriculture as a major component; they also include wider aspects of forest management and exploitation.

Artificial Intelligence - the branch of computer science directed to the solution of problems normally associated with human intelligence. The range of the subject is very large and a wide variety of techniques have been developed for specific application areas, which include trying to make computers understand natural human language, recognise and interpret visual scenes, capture the knowledge and experience of human experts and then apply them to solving problems (these are called expert or knowledge-based systems), or learn from experience (one technique uses neural networks which mimic very crudely the way the brain is thought to work). Techniques in artificial intelligence led to the development of windowing interfaces found on personal computers. Artificial intelligence is a component of many computer science courses but can also be studied in specialist degrees. Artificial intelligence has links with psychology and linguistics.

Astronomy/Space Science - the scientific study of the universe and the matter it contains such as the planets, stars and galaxies, the interstellar and intergalactic medium, comets, pulsars, quasars and black holes. Branches of the subject include astrophysics and cosmology (the study of the universe as a whole from its birth in the big bang, through its evolution to its current state, to a variety of conjectured futures). The basis of the subject lies in physics and mathematics. A wide range of observational techniques are used covering the spectrum from radio waves to gamma-rays, with electronic instrumentation and computer interpretation of data playing a major role.

Astronomy is taught as a topic in physics degree courses and in specialist degrees. The astrophysics component of astronomy degree courses is often stressed by using the title 'astronomy and astrophysics'.

Astrophysics - the branch of astronomy concerned with using principles from physics to explain processes taking place in the universe. Examples of work done in astrophysics include explaining the production of energy in stars, the evolution of stars over their lifetime, the behaviour of pulsars and the properties of quasars.

Biochemistry - the study of the chemistry of living organisms. The subject covers a very wide range of activity, including investigations into how an organism's metabolism is controlled and energy is supplied to the organism, the way the nervous system and brain operate, the action of muscles, the role of hormones and how they control body function, and the operation of the immune system. A wide range of techniques is used to investigate biochemical reactions, both within the organism and in standard conditions in the laboratory. Techniques are also used to separate, purify, synthesise and modify biomolecules. An important use of the last of these is in genetic engineering, where DNA is manipulated to alter the genetic properties of the organism.

Biochemistry has major applications in medicine, agriculture and, particularly since the development of biotechnology, a wide range of other industries including pharmaceuticals and the food industry.

Biochemistry is also an important component in biology, chemistry, medicine, veterinary science, dietetics, food science, pharmacy and agricultural science degree courses.

Biology (Biological Science) - the study of living organisms. Biology and biological science degree courses cover a wide range of topics, including plant and animal biology (botany and zoology), biochemistry, cell and molecular biology, genetics and ecology. The vast range of the biological sciences combined with the phenomenal progress that has been made in some areas mean that it is impossible to cover all of the subject in depth. Therefore, the later stages of most courses allow specialisation in one or more of the branches just mentioned, or in more specialised topics such as pharmacology, physiology, immunology, toxicology or microbiology. Work in areas such as biotechnology has recently opened up new potentials for the industrial application of biology. Greater awareness of environmental concerns has focused interest on the results produced by the study of ecology. Biology has links with agriculture, medicine, veterinary science, food science and environmental science.

Biomedical Science - these courses cover the scientific basis of medicine and draw on many of the biological sciences. The subject is interdisciplinary, combining work in the biological and medical sciences. Courses usually begin by providing a basis in biochemistry, cell and molecular biology and physiology. They build on this with more specialised study in areas such as pharmacology, genetics, nutrition, microbiology, immunology, pathology, neuroscience and even biotechnology in some cases, as well as courses more specifically related to work in hospital medical laboratories, such as medical biochemistry (for the analysis of body fluids), haematology (the study of blood), histology (the study of cells, particularly for cancer diagnosis) and immunology (for detecting antibodies and tissue typing). These courses are a good preparation for work in the research laboratories of health and biologically related industries. Some courses are also accredited by the Institute of Biomedical Science and the training they give may count towards the requirements for a professional qualification for working in the hospital laboratory service.

Biomedical science has links with biology, biochemistry and medicine.

Biophysics - the study of living organisms using ideas and methods drawn originally from physics. Work in biophysics includes investigations at the macroscopic scale, such as the mechanics of skeletal and muscle action, but courses now concentrate much more on the cellular and molecular levels. For example, electronic instrumentation and computer data logging may be used to investigate the cell membrane. At the molecular level, X-ray crystallography, ultraviolet and infra-red spectroscopy and nuclear magnetic resonance are used to investigate and determine three-dimensional molecular structures. Some of these techniques have been adapted for medical use through, for example, computer-aided tomography (CAT scanners). Specialised medical biophysics courses are available. The courses are generally taught in association with other biology courses; the physics content is usually rather less than the biology content and is directed specifically towards biological applications.

Biotechnology - the application of biochemistry, molecular genetics and microbiology to develop industrial processes based on biological activity. Biotechnology is an interdisciplinary subject drawing on results from the biological sciences (particularly biochemistry, microbiology and genetics), chemistry and chemical engineering, so courses cover aspects of all these, though the biological sciences tend to be the largest component. Biotechnology is a rapidly expanding area with great potential for many new and exciting applications in, for example, pharmaceuticals and medicine through the development of new drugs and medical treatments. It is now used in the chemicals industry, in agriculture through the use of genetic engineering to produce improved crops and livestock, and in the food industry, where brewing, for instance, represents an application of biotechnology that existed long before it became established as a separate discipline.

Botany (Plant/Crop Science) - the study of plants or in some cases crops. Although botanists have traditionally been concerned primarily with the discovery and classification of new plants, the emphasis is now much more on general principles of cell and molecular biology, plant physiology, ecology and conservation, though the emphasis given to each of these varies from course to course.

Crop science courses are also concerned with practical issues affecting cultivation and protection from disease and pests. Despite the fact that the subject is one of the oldest sciences, new techniques such as cell culture and gene cloning mean that it continues to offer fresh and exciting challenges.

Botany has links with all the other biological sciences, and with agriculture and horticulture.

Cartography - see mapping science

Cell Biology - the study of biology at the level of the cell, the fundamental unit of living organisms. Courses include work in biochemistry and molecular biology, common to nearly all biological science courses. However, they look in much greater detail at cell structure and function, cell membranes, the control, integration and behaviour of cells in multicellular organisms, genetics at the cellular level (cytogenetics) and a variety of techniques such as cell culture.

Ceramics Science - the study of ceramics. These are hard strong materials produced by firing mixtures containing clay. Familiar examples of ceramic materials are pottery, bricks and glazed tiles, but there are many other ceramic materials that are designed to have specific electrical, magnetic and heat-resisting properties that give them a wide variety of industrial and other applications. A well-known example of the use of specialist ceramics is in the space shuttle's re-entry heat shield.

Ceramic science is normally taught as part of a materials science or engineering course, but is also available as a separate specialist degree course.

Chemistry - the study of the properties and reactions of the elements and their compounds. Although the traditional division of chemistry into physical, organic and inorganic tends to be reflected in the organisation of the initial stages of courses, the boundaries between them are not sharp. Specialised options, usually offered later in the

Chapter 10 - Where do you go from here?

courses, do not always fit neatly into this classification. The applications of chemistry are so broad and varied that chemists can be found working in nearly all parts of industry, including not only areas such as chemicals and pharmaceuticals, but also rather less obvious ones such as the food industry and wherever materials are produced or processed (for example, in the steel and ceramics industries).

Chemists also work in the public sector in analytical laboratories monitoring health and safety. They are at the forefront in the fight against environmental problems; for example, it is chemists who have to find effective alternatives to environmentally harmful substances such as CFCs or pesticides.

Chemistry courses usually provide opportunities to specialise late in the course. However, there are also more specialised degrees, such as biological, medicinal and environmental chemistry, which concentrate on one particular area, though they usually build from a solid foundation in general chemistry.

Chiropody - see podiatry

Computer Science/Studies - the study of the principles and use of computers. Courses vary greatly, some taking a scientific/mathematical approach, while others concentrate on the practical uses of computers. Many courses have a hardware component covering basic logic circuit design and computer architectures, though more specialised topics such as VLSI (very large scale integrated circuits) design may also be available as options. The purpose of the hardware component is usually to give a background for other parts of the course. Courses in electronics or electronic engineering may be more suitable for a professional training as a hardware engineer. The largest amount of time is usually devoted to work on software. All courses cover at least one programming language, and some considerably more, as well as operating systems (the programs used by computers to control their hardware and to run the user's programs), though the depth to which the underlying principles of operating systems are covered will vary. The greatest variation between courses comes from the way they treat applications software. Some concentrate on the low-level software engineering aspects of producing applications, others more on their use in particular areas such as for business information systems.

Artificial intelligence is a major area of computer science, which some courses cover in considerable depth. They may cover both the basic theory, some of which is drawn from psychology and linguistics, and the use of specific techniques such as expert systems (also called knowledge-based systems) or neural networks.

Courses vary considerably in how much formal theory they include and the depth to which they go. The theory has been developed in an attempt to help programmers produce 'correct' programs (in the sense that they behave as intended in all circumstances). The need for this arises from the fact that only very small programs can be tested exhaustively. The theory uses ideas and notation drawn from mathematics; in its more advanced forms it is highly mathematical.

Courses in the area of computer science are available under a number of other titles, such as 'computing', 'information technology' and 'business computer systems', which usually reflect a more specialised orientation. Many other subjects include topics in the use of computers, though they are usually only concerned with the use of methods and applications relevant to that subject.

Crop Science - see botany

Cybernetics - the study of systems and how they are controlled. Applications include industrial automation and robotics. Techniques are drawn from a wide area including electronics, computer science, artificial intelligence and instrumentation.

Dentistry - dentists work to conserve their patients' teeth and to treat problems of decay, gum disease and misalignment if they arise. The courses last five years and involve considerable clinical practice. The basic content of all courses is very similar, but they differ in the way it is organised, the emphasis given to different elements and the possibilities for elective studies in the later stages of the course.

Dietetics - the scientific study of diet and nutrition. Courses with 'dietetics' in their title usually provide a period of professional training in hospitals, and can lead to state registration, which is required for professional practice as a dietician in the National Health Service. Dieticians also work in the food industry and in private practice. Courses include components of physiology, biochemistry and microbiology, together with elements drawn from the behavioural and social sciences.

Earth Science - an interdisciplinary subject combining the study of geology (the largest component) with aspects of physical geography and, in some cases, oceanography, meteorology and climatology.

Chapter 10 - Where do you go from here?

Ecology - the study of the interrelationships between plants and animals, and between them and their environment. It looks at how the environment influences and is influenced by individual organisms, populations of an individual species and communities of different species. Ecology is normally taught alongside other biological science courses and often shares with them a basic grounding in topics, such as plant and animal biology, cell and molecular biology and genetics. Issues of conservation and environmental damage through pollution and other factors are usually covered. The later stages of courses often allow some specialisation in the study of specific ecosystems, such as marine, freshwater, or agricultural ecology.

As well as being available as a specialist degree subject, ecology is taught in many biological science courses. It is also a major component in environmental science courses with a biological science orientation; specialised ecology courses generally give less emphasis to social and policy issues.

Electronics - the study of the construction and use of electronic devices. These devices are now almost exclusively semiconductor-based. Courses usually give a balanced treatment of both linear (analogue) and digital electronics, though some more specialised courses are available. A wide range of applications is normally covered from areas such as instrumentation, control systems, computer design and architecture, and audio-electronics.

Electronics is available as a specialist topic within other subjects such as physics and mechanical engineering.

Environmental Health - the courses are concerned with the health implications of environmental factors such as food safety, pollution and noise. Courses cover the relevant basic science, such as microbiology, anatomy and environmental studies, as well as more specific topics, such as food safety and hygiene, and occupational health and safety. They also cover the relevant legislation and how it is enforced. Qualified environmental health officers work mainly for local authorities to improve standards of health and safety, particularly in food production and service, but also in other aspects of people's living and working environments.

Environmental Science/Studies - an interdisciplinary subject studying the environment and the very wide range of factors affecting it. The content and emphasis of courses varies considerably. Some courses have a physical sciences (most particularly, chemistry) or earth science orientation; rather more have a biological sciences bias and others a balance of approaches, though in the latter case options may allow some specialisation in a particular area. Many of the courses with a biological orientation may share appreciable content with courses in ecology, but tend to be more concerned with practical and social issues, and the resulting consequences for policy. This means that many courses also include relevant aspects of law, economics and other social sciences. There is a wide variety of courses specialising in specific aspects of the subject, such as environmental management, environmental chemistry, environmental biology or environmental toxicology (pollution).

Ergonomics - the study of human beings in their working environment, and in particular the design of working practices, equipment and tools for optimum efficiency and protection of workers' health and safety. Courses include relevant aspects of anatomy, psychology, physiology, industrial engineering and design. Issues dealt with include furniture design, factory and office layout, and environmental conditions such as heating and lighting.

Food Science/Studies - the scientific study of food, including production, processing, storage and distribution as well as nutrition. Food science courses are firmly based on chemistry and biochemistry, with contributions from microbiology and physiology. Food studies courses usually have a less intensive science content and give more emphasis to the general aspects of food such as related economic and social issues. Both types of course include topics relevant to management in the food industry, though there are also specialised courses dealing with business aspects of the food industry, such as food marketing, which often include some work on food science.

Forestry - courses cover all aspects of the cultivation, exploitation and management of forests. Building on a foundation of basic science including chemistry, geology, botany, zoology and soil science, they cover more specialised topics such as wood structure, tree and forest husbandry (silviculture), tree pests and diseases, forest measurement, land use and urban forestry, together with economics and management. There is a strong emphasis on practical work and prominence is given to commercial as well as technical aspects.

Genetics - the study of the inheritable characteristics of organisms. Genetics evolved through the study of the development of genetic variation in whole populations, sometimes over long periods.

Although this still plays a part, the emphasis of courses is now much more at the cellular and molecular level, where an understanding of the structure of DNA and the way it can be manipulated allows the possibility of changing the genetic make-up of organisms through the techniques of genetic engineering. The scientific, medical and commercial applications of these techniques are immense and only just beginning to be realised. Courses in human genetics often cover the topic of genetic counselling, which requires technical, social and personal skills.

Genetics courses are often taught in parallel with other biological science courses, with much of the content shared in the early stages. Genetics is also taught as part of courses in other biological sciences, medicine, veterinary science, agriculture and horticulture.

Geochemistry - the chemistry of rocks and the processes occurring during rock formation and transformation. Chemistry is an important tool for the geologist, providing both a method of identification through analysis and the means for explaining many of the properties of rocks and the processes taking place during their formation. Geochemistry is therefore often an important component of geology courses, but specialist courses are also available allowing for deeper treatment of the subject. Geochemical prospecting is used commercially to find deposits of ores from analyses of soils, water courses and sediments.

Geological Science - courses combine work in geology, geophysics, geochemistry and applied geology.

Geology - the study of the Earth, including its composition, structure and historical development from its creation five billion years ago to the present day. Geologists study the rocks appearing at the surface of the Earth, but also use less direct evidence such as the way seismic waves are transmitted within the Earth. The theory of plate tectonics has had a major influence on geology in recent decades and is used as a unifying concept in many courses. Geology courses draw on a wide range of basic science and cover many different topics including crystallography, mineralogy (the study of the crystalline material that forms rocks) and petrology (the study of rock origin, composition, structure and alteration), geochemistry, geophysics, stratigraphy (the study of strata laid down over time) and palaeontology (the study of fossils). The applications of geology to oil and mineral prospecting and extraction are obvious, but it is also important for civil engineering projects, such as tunnel boring and the building of dams and reservoirs. A knowledge of the geology of the area is also important for the analysis of environmental problems, such as the flow of pollutants from ground water to water courses or drinking water.

Geology is a major component of earth science courses and some courses in environmental science.

Geophysics - the study of the physical properties of the Earth (and by extension, other planets). By studying these properties, geophysicists are able to deduce the structure of the Earth and the nature of processes within it at great depths beneath the surface. Geophysicists draw on and develop a wide range of physics, including theories of fluid and heat flow, electromagnetism, gravitation and wave propagation through solids and fluids. Data are gathered using sophisticated electronic instrumentation, including instruments based on satellites and space probes. Computers are used extensively to control instruments, process the data and then interpret it through the use of complex mathematical models. Geophysicists make a major contribution to the search for oil and minerals. They are also involved in work on earthquakes, which may in time enable them to give reliable predictions of when and where they will occur.

Geophysics topics appear in most geology and some physics courses.

Health Studies - covers a variety of different courses where biological and medical subjects are combined with social studies, health administration and in some courses sports management and administration. The courses are often part of combined or modular degrees with a wide choice of options.

Horticulture - the cultivation of plants for food (commercial horticulture) or to enhance the environment (amenity horticulture). Courses start with a foundation in basic science usually including botany/plant science, biochemistry, soil science, ecology and genetics. They then introduce more specialist horticultural topics such as plant production, plant pests and diseases, crop protection, landscape management and the use of organic methods. All courses place strong emphasis on practical work and prominence is given to commercial as well as technical aspects throughout. There is usually a wide range of options in the final year and it is often possible to specialise in commercial or amenity horticulture at that stage.

Horticulture has links with the biological sciences, forestry and agriculture.

Immunology - the study of the system used by organisms as a protection against infection. The function of the immune system has been recognised for many years and this enabled the discovery of vaccines. More recently, however, there has been growing interest in the immune system and problems connected with it. One reason for this was the development of transplant surgery, where the immune system must be counteracted to prevent the rejection of the transplanted tissue. A class of diseases called auto-immune diseases, which includes rheumatoid arthritis, has also been recognised. In these, the immune system malfunctions and acts against the body's own tissues. There are also conditions and diseases, such as AIDS, where the immune system breaks down and leaves the body unprotected against infection.

The courses are closely linked with the biological sciences and medicine, and develop from a foundation including biochemistry, physiology and cell and molecular biology. Immunology can be studied in a specialist degree course or as part of a course in biological science, medicine or veterinary science.

Information Science/Management - the study of the collection, storage, retrieval and dissemination of information. Information science courses provide professional qualifications in librarianship. The courses contain topics such as classification systems that are directed specifically to library work, but much attention is also directed towards the creation and use of computer-based information systems. This means that the courses are now less distinct from other courses directed at the information industry, such as information management, information systems and information technology.

Information Technology - this title is used for two rather different types of computing course. The first type has an engineering orientation with rather more emphasis on hardware, particularly in connection with communications and networks, than is usual in computer science courses. The second type of course is concerned mainly with the use of business or management software systems, with less emphasis on low-level programming and hardware than in most computer science courses. Courses with titles such as business information technology are usually of this second type, but in general the only way of telling the orientation of a specific course is from a detailed description of its contents.

Linguistics - the scientific study of language and its structure. This includes the way sounds are used to make speech (phonetics and phonology), the way sentences are constructed (syntax), the way meaning is conveyed using words and sentences (semantics) and the way language is used in context (pragmatics).

Linguistics is also a component of some courses in foreign languages, speech science, education, computer science (particularly artificial intelligence) and psychology.

Mapping Science (Cartography) - the study of the processes required for the creation of maps. The process of creating a map can be divided into data acquisition, analysis and presentation. One aspect of data acquisition is surveying, which includes land-based and remote sensing using aerial and satellite imaging. However, maps are used to display a wide range of information, including such things as market research results, so data acquisition also includes issues such as social survey and sampling techniques. The analysis phase includes image analysis and processing, as well as statistical analysis. The presentation stage covers a wide range from manual mapping to the construction of computer-based geographical information systems.

Mapping science is often a component of degrees in surveying, geography and geology.

Marine Biology - the study of the biological systems found in the sea, including individual organisms and a variety of ecosystems. The courses build from a solid foundation in biological science, covering basic plant and animal biology, cell biology, biochemistry, genetics and evolution. An understanding of the marine environment is gained through the study of physical, chemical and atmospheric processes within and around the sea and oceans. This work is built on with more specialised courses in topics such as the physiology of marine organisms, food chain processes, behaviour, marine pollution, conservation, fisheries and aquaculture.

Marine biology is also available in combination with freshwater biology and as a specialised option in some biology/biological science courses.

Materials Science - the study of the physical properties of natural and man-made materials used in industry and construction. Many of the advances that have led to an increase in the standard of living this century have been brought about by a greater understanding of the way materials can be used and by the introduction of new materials that have improved properties and are cheaper to produce. For example, much of the engineering industry relies on the use of high-performance alloys. Courses cover a wide range of materials, which may include natural materials such as wood or stone, though the emphasis is more commonly on artificial or processed materials such as ceramics, metals, semiconductors and plastics. The subject is very active and promises exciting developments; for example, the production of new semiconductors offering greatly enhanced computer performance and the recent development of high-temperature superconductors. The courses draw from a variety of other subjects but are principally based on solid-state physics and chemistry.

Some courses allow the specialised study of individual materials later in the course and others from the beginning. Materials science is also a major part of materials engineering degree courses and is often a component of mechanical engineering courses.

Mathematics - a broad and diverse subject with a history dating back to 3000 BC, but one that is developing more rapidly now than at any time in the past. There are some broad divisions within the subject, which you can concentrate on by taking a specialist degree or by taking options in a general mathematics degree. Pure mathematics includes topics such as calculus, algebra and geometry that have familiar titles, though the content of the last two in particular is markedly different from what you have experienced at school. It also includes topics that are less familiar, such as logic, ring and field theory, number theory and topology. Applied mathematics includes topics such as mechanics, electromagnetic field theory, fluid mechanics, relativity and quantum theory and elementary particle physics. Applicable mathematics includes a variety of techniques that are used widely in business and covers topics in statistics and methods of optimisation such as linear programming. Statistics is also a component of many mathematics courses.

Virtually all courses include some computer science. There are many specialist mathematics courses dealing with particular application areas, such as business mathematics or mathematical physics. All science and engineering courses include some mathematics.

Medical Physics - the study of the interaction between the body and all types of radiation used for diagnostic and therapeutic purposes. Diagnostic techniques include X-rays, ultrasound scans, gamma-ray imaging and magnetic resonance imaging. Therapeutic techniques include laser surgery and radiation treatment. Courses start with a foundation of mathematics and physics, including electromagnetism, atomic physics, nuclear physics and optical physics. This is built on with more specialised topics such as biophysics, physiological medicine and radiography.

Medical physics options are available in the later stages of several physics courses. There are also medical electronics courses, which concentrate on physiological monitoring and instrumentation.

Medicine - courses leading to professional qualification as doctors. The courses last five or six years. They are based on a foundation in basic medical science subjects including human anatomy and physiology, biochemistry, microbiology, pharmacology, genetics, immunology and pathology. The remainder of the course is devoted to clinical training, though some formal teaching continues throughout the course. Able students may be allowed to extend their studies by spending an extra year in the middle of the course studying some aspect of medical science in greater depth. These 'intercalated years' lead to the award of an additional degree. Courses usually contain an elective period towards the end when students can broaden their experience by performing a research project or gaining clinical experience in a new environment such as in a foreign country. At the end of the course, a further year of experience leads to registration with the General Medical Council, but several years' further training is required before appointment to a permanent post in hospital medicine or general practice. There are also other courses allied to medicine which lead directly to professions such as chiropractic, osteopathy, and prosthetics and orthotics.

Metallurgy - the study of the extraction, purification, production and properties of metals. Metals play a vital part in all areas of industry, especially in areas such as engineering and construction. Metallurgists ensure that metals have the correct properties to perform the task required effectively and safely. They create high-performance alloys for special purposes in industries such as aerospace, as well as ensuring, for instance, that the quality of steel used in car manufacturing is consistent. How metals are affected by exposure to the elements depends on their properties, which in turn depend on their constituents and the way they are processed. Metallurgy courses draw on material from a wide range of other subjects including chemistry, physics, geology and crystallography in order to understand these relationships.

Metallurgy is also available as part of materials science/engineering courses, and has links with mechanical and other engineering degrees.

Meteorology - the scientific study of the weather and atmospheric processes. The behaviour of the atmosphere is extremely complex and meteorologists have to draw on a wide range of information, techniques and theories. Courses include physics and mathematics and topics such as atmospheric dynamics, atmospheric physics, climate change, surface processes and oceanography. The meteorologist uses acquired knowledge to build sophisticated mathematical models and computer simulations to understand and predict evolving weather patterns.

Meteorology is available as a single honours course, but is more frequently combined with physics or taken as part of a course in environmental or earth science.

Microbiology - the scientific study of micro-organisms including bacteria, viruses, protozoa and some algae and fungi considered small enough to qualify. The courses are often taught alongside other biological science courses and often share with them a basic grounding in topics such as plant and animal biology, biochemistry, cell and molecular biology, genetics and ecology. This leads to more specialised topics such as bacteriology, virology, mycology (the study of fungi) and immunology. Courses often have a range of options related to the application of microbiology to agriculture and medicine. The importance of microbes as agents of disease has long been recognised, but the explosion of interest in biotechnology in recent years has given extra impetus to the study of microbiology. For example, bacteria can be genetically engineered so that they produce human enzymes or hormones. The bacteria can then be grown on an industrial scale to produce these compounds commercially for medical use.

Microbiology is also available as a component of degrees in biological sciences, medicine, veterinary science, dentistry, horticulture and agriculture.

Molecular Biology - the study of the structure and reactions of large biological molecules such as nucleic acids and proteins. The courses are often taught alongside other biological science courses and often share with them a basic grounding in topics such as chemistry, biochemistry, plant and animal biology, cell biology and genetics. Other topics that may be taught later in the course include microbiology, immunology and biotechnology. Molecular biology is at the centre of many of the most exciting developments in biology and its influence has spread to many other branches of the subject, such as genetics. The scientific, medical and commercial applications of genetic engineering have created even greater interest in the subject.

Molecular biology is often combined with cell biology and is also available as a component of degrees in biological science, biochemistry and agriculture.

Neuroscience - the study of the nervous system including the brain. Neuroscience draws on a wide range of other biological and medical subjects including molecular biology, cell biology (to study the chemical and electrical communications between cells), and psychology (to study the behaviour of the whole organism). Because of this, the courses normally begin with a solid foundation of biochemistry, animal biology, cell and molecular biology, pharmacology and physiology. Understanding the function of the brain is one of the most challenging problems in modern biology and medicine, with many potential applications.

Neuroscience is available as a single honours course, but is more frequently taken as an option towards the end of physiology courses.

Nursing Studies - degree courses that include nursing training but enhance it to a degree-level. Some courses extend the depth of nursing subjects and are based in science/medical faculties. Others extend the breadth and offer more social sciences such as sociology, psychology, economics, health policy and health studies. Some have a balance of these approaches. All courses will cover anatomy, physiology, pharmacology, human development, nursing techniques and practical training, and most include some management studies. Degree courses begin with an 18-month Common Foundation programme followed by a branch programme, specialising in the adult, mental health, learning disability or children's branch of nursing. Most institutions offer only one or two of the branches. There are also associated degrees in nursing where students can take a degree at a university in one or more subjects and take a nursing course in a nearby school of nursing. There are some one- or two-year degree courses for students who are already qualified nurses.

Nutrition - the study of the body's requirements for food and how they can be met. The courses are often similar to and taught with courses in dietetics, except that they do not have a period of professional practice. There are also a number of courses in animal nutrition, which are taught alongside agricultural sciences.

Nutrition can also be studied as part of courses in physiology, food science and agriculture.

Occupational Health and Safety - the study of health and safety at work and in the wider environment, and how it can be maintained and improved. Subjects studied include biochemistry, physiology, ergonomics, occupational health, environmental measurement methods, toxicology, law and risk management. These courses are specialised training for jobs in health and safety advice and inspection.

Occupational Therapy - the courses provide professional training for occupational therapists. Qualified occupational therapists are employed in hospitals to help people overcome the effects of mental or physical illnesses or disability so that they can live as full and independent lives as possible. The courses cover human biology, psychology, anatomy and physiology, as well as the principles and techniques of treatment. All the courses include practical work with patients suffering from physical and/or psychological difficulties or disabilities, and draw on personal and social skills as well as technical knowledge.

Oceanography - the scientific study of the oceans. Oceanography is a very broad subject drawing on many other disciplines. For example, understanding the intricate patterns of ocean currents, tides and waves requires complex computer models drawing on theories of fluid flow from physics and mathematics. Studies of the sea-floor apply knowledge from geology. Marine chemists study interactions at the interfaces between sea-water and materials in the seabed, and sea-water and the atmosphere at the surface. Marine biologists investigate all aspects of life in the sea. The practical importance of understanding the oceans is immense. For instance, the oceans have been used to absorb vast amounts of waste material, and only with a proper understanding of processes within the ocean can the resulting risks be assessed. The oceans are a major food resource and support major transport systems. Oil is already being taken from below the seabed and there is potential for extracting other materials. The oceans also have a major effect on climate.

Oceanography can also be studied as part of a course in earth science or geophysics.

Optometry (Ophthalmic Optics) - the courses provide professional training for optometrists (also called ophthalmic opticians). Optometrists carry out eye tests to detect abnormalities and vision defects, analyse the results, and prescribe and dispense spectacles and contact lenses.

Optometrists also carry out work in orthoptics and check for diseases of the eye such as glaucoma, referring patients to specialist doctors (ophthalmic surgeons) for treatment. The courses include the anatomy and physiology of the eye, optics, clinical optometry (measurement), methods of examination and diseases of the eye, as well as clinical work with patients. They also draw on personal and social skills and technical knowledge.

Orthoptics - the diagnosis and treatment of disorders of the eyes caused by problems with the muscles controlling the eye. The courses provide professional training in orthoptics. Much of the orthoptist's work is with young children, dealing with squints, for example, but they also work with adults suffering from a variety of conditions such as the after-effects of accidents and strokes. Orthoptists can treat some of the problems they diagnose, but refer more severe cases to specialist doctors (ophthalmic surgeons). The courses include the anatomy and physiology of the eye, optics, child development and ophthalmology (the scientific study of the eye). The courses also include clinical work with patients and draw on personal and social skills as well as technical knowledge.

Pathology - the study of disease and disease-causing micro-organisms such as bacteria and viruses. It includes the study of microscopic and macroscopic changes in the organisation of tissues, and chemical changes in tissues and body fluids. The courses start at the study of healthy organisms and progress to contrasting them with those affected by disease. Anatomy, physiology, cell biology, genetics and microbiology, as well as more specialised topics such as cell pathology and haematology (the study of cells circulating in the blood-stream), are included on courses.

Pathology is available as a single honours course, but is more frequently taken as a component of courses in biological and biomedical science, medicine, veterinary science and agricultural science. To become a professional pathologist, you must first qualify as a doctor.

Pharmacology - the study of the action of drugs on living systems. Courses build from a foundation of chemistry, biochemistry, cell and molecular biology, physiology and pathology. In addition, there are specialist topics within pharmacology such as the mechanisms of drug action and drug toxicity. Pharmacologists are involved in medical research and play a vital role in the pharmaceuticals industry helping with the development and screening of new drugs for use in medicine, veterinary science and agriculture.

Pharmacology is also studied as a component in some agriculture, biochemistry and biological science courses, and is an integral part of the training for pharmacy, medicine, veterinary science, dentistry and nursing. Indeed, many professional pharmacologists originally trained in medicine or pharmacy.

Pharmacy - the courses provide professional training for pharmacists. About one third of the course is spent studying pharmacology. The rest of the time is used to study basic subjects such as chemistry and microbiology, as well as more specialised topics such as pharmaceutical chemistry (the structural analysis and synthesis of drugs), pharmaceutics (the manufacture of drugs and their formulation into a suitable form for humans or animals) and clinical pharmacology (the study of the effects of drugs on the human body). Other subjects that may be studied include toxicology (the study of poisons) and pharmaceutical engineering.

Physics - the science concerned with the behaviour and properties of matter and energy, but largely excluding those processes involving a change in chemical composition. Physics encompasses investigations at the smallest possible scale of sub-elementary particles called quarks, as well as explanations of the origin and evolution of the entire universe using general relativity. Indeed, attempts are now being made to explain both these extremes within a single unified theory. However, even if such a theory is produced, there will still be many challenging fundamental problems left for the physicist in areas such as low-temperature physics, solid-state physics, plasma physics and astrophysics to name just a few, as well as in other areas where the techniques of physics are used, such as medical physics and geophysics.

Courses usually build from a solid foundation of classical physics covering topics such as mechanics, electromagnetism, properties of matter, thermodynamics and statistical mechanics. To these are added quantum mechanics and relativity, which were developed in the early years of this century and revolutionised the way physicists and others viewed the world. They are still the two key theories of modern physics. In the later stages of courses, more specialised topics can usually be studied, such as nuclear physics, astronomy, astrophysics, biophysics, geophysics, X-ray crystallography, medical physics, elementary particle physics, electronics and laser physics. Mathematics is also an important component of all physics courses, particularly those with options in theoretical physics.

Physics topics are also studied within all engineering, electronics and applied mathematics courses.

Physiology - the study of how animals function (plant physiology appears as a title for topics in botany and biological science courses, but you can assume that courses called physiology are only concerned with animal, and often just vertebrate, physiology). Physiology covers all levels, from the subcellular, such as the operation of cell membranes, through the function of individual organs, such as the way the heart operates as a pump, up to the way the animal functions as a whole. In explaining how organisms function, physiology draws on ideas and techniques from a wide range of other subjects including physics and chemistry, and many of the biological sciences such as anatomy, biochemistry, genetics, pharmacology and microbiology.

Physiology can also be studied within courses in biological science, human biology and agriculture, and is an important component of courses in medicine, dentistry, veterinary science and most of the professions allied to medicine such as physiotherapy and radiography.

Physiotherapy - the courses provide professional training for physiotherapists. Qualified physiotherapists work in hospitals and the community. They use physical methods including manipulation, massage, infra-red heat treatment and remedial exercises to treat clients of all ages who may be suffering from any of a wide range of disabilities and disorders, such as arthritis, mental illness, stroke

Chapter 10 - Where do you go from here?

and accidents. Courses include anatomy, physiology, pathology, human science, biomechanics and treatment techniques. All the courses include clinical work with clients and draw on personal and social skills as well as technical knowledge.

Plant Science - see botany

Podiatry (Chiropody) - the diagnosis, treatment and management of conditions affecting the foot and lower limb. Courses provide professional training that can lead to state registration, which is required for working in the National Health Service. Clinical practice is an important part of all courses and can occupy up to half of the time. Theoretical work covers the structure and function of the normal limb, the biomechanics of locomotion and gait, pathology, podiatric medicine, behavioural science and treatment techniques. Qualified podiatrists work in hospitals and the community, in both the National Health Service and private practice.

Polymer Science - the study of polymers (molecules made up of long chains of repeating units), including their formation, properties and uses. Polymeric materials come in many different forms such as plastics, fibres and foams; they may be synthetic, such as nylon or polystyrene, or natural such as rubber or cellulose. It is possible to fine tune the properties of polymers in the manufacturing process, so they can often be adapted for different applications. They can also be combined with other materials to form composites that display the best properties of each material. The courses are specialised materials science courses. They are often taught alongside other materials science courses and share some common content.

Polymer science can also be studied as a component of courses in materials science and chemistry, though the approach in the latter will be rather different.

Psychology - the scientific study of the mind. Psychology is a very broad subject and different courses may emphasise very different aspects of the subject. One reflection of this is that it may be offered as a BSc within a science or social science faculty or a BA within an arts faculty, though if they are offered at the same institution much of the teaching may be common with the main differences being the options offered. Courses also vary in the amount of work on animals, as well as in the emphasis given to different branches of psychology such as experimental psychology, cognitive psychology (the study of thought processes, memory and language), physiological psychology (the study of the relationship between brain function and psychological processes), comparative psychology (the study of animal behaviour and the similarities and differences between species), developmental psychology (changes in psychological processes during maturation into adulthood), social psychology (how people and animals behave in groups), the psychology of individual differences (the study of factors such as personality and intelligence), abnormal psychology (the study of abnormal behaviour) and applied psychology (the application of psychology to practical problems, for example in negotiation strategies).

Psychology can also be studied in a wide variety of other courses including education, medicine and the professions allied to medicine such as nursing and occupational therapy.

Radiography - the courses provide professional training in diagnostic or therapeutic radiography. The two branches of the profession are distinct and the courses separate, though some teaching may be common to both. Diagnostic radiographers work in the X-ray departments of hospitals or in health centres. They use a wide variety of techniques, including X-rays, ultrasound, magnetic resonance imaging and computer-aided tomography to produce images to help doctors perform diagnoses. Therapeutic radiographers work in hospital clinical oncology (cancer) departments. They plan and administer treatment using a variety of radiation sources such as X-rays, isotopes and linear accelerators. Both types of course include anatomy, physiology, pathology, radiographic physics, radiographic techniques, patient care and professional practice as radiographers. All the courses include clinical work with patients and draw on personal and social skills as well as technical knowledge.

Speech Sciences/Speech Therapy - the study of speech, including anatomy and physiology of the mouth and throat, phonetics (vocal sounds), linguistics (the structure of language), psychology and the pathology of speech problems. Other topics covered include acoustics, audiology, neurology and education. Speech therapy courses also include clinical practice and contribute to a professional qualification in the diagnoses and treatment of speech defects. Speech therapists work with children in schools and clinics, and with adults with speech problems, such as stroke patients.

Sports Science - the study of human movement, physiology, anatomy and psychology in relation to sporting activities. Many courses also include topics on the management of sports facilities.

Chapter 10 - Where do you go from here?

Statistics - the science of collecting and analysing numerical data. Statistics is a major branch of mathematics with applications in practically every area of scientific, commercial and administrative activity. For example, the results of any scientific experiment must be shown to be statistically valid before they are accepted, the demand for new products has to be assessed before they are launched, and the provision of education and health services has to be planned in advance to cope with changes in population and technology.

Courses cover a range of basic mathematical subjects such as calculus and algebra, as well as more specialised topics such as probability analysis, statistical distributions and a wide range of statistical analysis techniques. The use of computers is essential for the statistician, so courses include numerical analysis and computer science, with particular emphasis on the use of a wide range of software packages for statistical analysis in various fields.

Statistics can also be studied in combination with mathematics or as part of a mathematics degree. Topics in statistics are also taught in all science and medical degree courses and form important components of courses in the social sciences, such as psychology and business studies.

Veterinary Sciences - the courses provide professional training in veterinary medicine. Courses last five or six years and include many similar subjects to medical degrees, such as anatomy, biochemistry, physiology, medicine, surgery, obstetrics, pharmacology and pathology, though the focus is naturally on animals rather than humans. Courses cover large animal (farm livestock) and small animal (domestic pets) practice, as well as some exotic animals. Practical clinical training is a major component of all courses and dominates the second half of the course. It is usually carried out in clinics attached to the veterinary school and by attachment to practising veterinarians. Experience of practical work with livestock on a farm is also part of the training.

Zoology (Animal Science) - the scientific study of animals including their anatomy, physiology, classification, distribution, behaviour and ecology. Courses often run alongside other courses in biological science and share with them a foundation in biochemistry, cell and molecular biology, microbiology and genetics. Many courses stress the practical importance of zoology, with specialist topics such as fisheries biology and parasitology. In fact, some animal science courses are directed specifically towards agricultural applications.

Zoology can also be studied as a component of courses in biological science, agriculture and environmental science.